When Communities Confront Corporations
Comparing Shell's Presence in Ireland and Nigeria

Published by
Adonis & Abbey Publishers Ltd
P.O. Box 43418
London
SE11 4XZ
http://www.adonis-abbey.com
Email: editor@adonis-abbey.com

First Edition, May 2008

ISBN: 9781906704049 (HB)

Printed and bound in Great Britain

When Communities Confront Corporations

Comparing Shell's Presence in Ireland and Nigeria

By

Austin Onuoha

Adonis & Abbey Publishers Ltd

Table of Contents

Chapter 3
Deductions

Chapter 4

Acknowledgement

When I returned from the US on a speaking tour on oil and poverty in Africa in November 2005, Rev. Fr. Kevin O'Hara handed to me another invitation to be a guest speaker at another conference in Dublin, Ireland. I accepted the invitation because the theme of the conference sounded unique to me. This was known as the Annual Feile Bride Conference in Kildare, Ireland. Feile Bride means the Festival of Brigid, who is the patroness of Ireland. She lived in the 6th century and is renowned for her commitment to justice, peace and equality. She lit a flame in Kildare, which burned for 1000 years until it was finally extinguished. But on February 1st 2006, the flame was permanently re-lit once again. This flame and the significance of flames in the oil industry became the theme of my address at the conference.

This conference which has been held for the past fourteen years has focused on different justice and peace issues. The theme of this conference is *Shining a Light on Exploitation: Human Rights Defenders.* The conference examined such issues as exploitation of women, exploitation caused by war and conflict and exploitation of resources by companies such as Shell.

During this conference I met and spoke with people from Erris. In fact the 'leader' of the *"Shell-to-sea Campaign"* Maura Harrington was on the same panel with me at the conference. Part of my itinerary during the conference was to visit Erris in Mayo County in the West of Ireland where a flame of conflict is burning. This issue which is now known as the *Rossport Five* involved the imprisonment of five men – Michael O'Seighin, Brendan Philbin, Willie Corduff, Vincent McGrath and Philip McGrath from the west of Ireland for 94 days for opposing Shell's attempt to force 'a dirty and dangerous' pipeline through their farms and close to their houses. While the men have been freed from prison, the issue remains unresolved.

I shall remain eternally grateful to Rev. Fr. Kevin O'Hara who facilitated my trip to Ireland both in 2001 and 2006. I am also grateful for his support and encouragement and more importantly for his vision in founding the Centre for Social and Corporate Responsibility (CSCR). Let me also thank Terry Conway, the man who drove me to Erris in his car and took me on a tour of the pipeline route. Terry's humility humbled me. I am also grateful to two of the *Rossport Five* Michael and his elegant wife Caitlin who hosted me for one night each in their home. They engaged me in lengthy conversations that clarified a lot of the Rossport issue to me. I am also grateful to Vincent McGrath and his lively wife Maureen. They hosted me in the serenity of their home on my second day in Erris. I loved the view of the mountain from their beautiful dining area. The effervescent innocence and freshness of their home illustrated in no uncertain terms the need for a clean and healthy environment.

I met Maura Harrington at the Annual Feile Bride Conference in Kildare, Ireland. In fact we sat on the same panel. While Maura spoke about Shell's presence in Erris, I spoke and compared what was happening in Erris with that of the Niger Delta. I thank Maura for speaking with me all through my stay in Ireland. Maura inspired me because of her commitment, courage and passion about Erris.

For the technical and analytical background of this book I am heavily indebted to the Centre for Public Inquiry for their report titled, "The Great Corrib Gas Controversy" written by Frank Connolly and Dr. Ronan Lynch. This report proved very helpful in corroborating data from the interviews which I conducted and from other sources. The independent analysis that was prepared by Accufacts Inc. for the Centre for Public Inquiry which is an appendix of the report which I referred to above was also an invaluable resource.

I met and spoke with so many other people who space may not allow me to acknowledge, but they know themselves. And I thank them. At the Centre for Social and Corporate

Responsibility (CSCR) in Nigeria, I wish to acknowledge everyone that has supported and shared my work. I am indeed very grateful to the staff at Warri office Florence, Nelly and Stanley who have made tremendous sacrifices in their first few months of work. And to my friends and professors at the Department of Conflict Analysis and Resolution of Nova Southeastern University in Florida I say thank you. And in this work and field I have engaged in intense debate with my friend and colleague – Paulinus Okoro, in spite of our disagreements we have always shared the same vision for our world.

Can I ever forget the good men and women at Misereor in Germany who gave me the grant for my studies in the USA. The opportunity offered to me by the management and staff of Misereor is one of the main reasons why this book did not die in my head. I am also grateful to the Conflict Transformation Program (now the Center for Justice and Peacebuilding) at the Eastern Mennonite University, Harrisonburg, Virginia for nurturing and nourishing me. I shall ever remain grateful to the people at EMU.

Finally, while I go about doing my work of trying to change the world, some people have paid heavily for my absence. I wish to thank my very beautiful wife Lynda for her love, support and steadfastness. And to my kids Iheabunike and Ndudi, I do hope that one day when you come of age you will be able to understand why I am never at home and forgive me for it.

Austin Onuoha
Africa Centre for Corporate Responsibility
Warri, Nigeria
June 2007.

Preface

I had my university education when pan-Africanism was the in-thing. During this period I was made to understand that the white man was the source and cause of all my problems. I was also told that *the only solution* (emphasis mine) was through a Marxian revolution. I assimilated all these and did all I can to either stimulate a revolution or waited for one to happen. When I graduated from the university, I was posted to Kontagora in present day Niger State of Nigeria to serve the nation.

Five of us corps members were housed in a large bungalow at the outskirts of the town. On the day we moved into the house, an Irish Catholic priest who lived next door invited us for a drink. While we were in his house, he asked me my name. I told him. He said that he recognised my name as someone from the eastern part of Nigeria. I agreed with him. He also said that he raised money for the people of the east when they fought the civil war against the Nigerian government in 1967.

At this revelation I was startled. And my only response was to ask him why he did it. He told me that they the Irish have experienced more than 300 years of colonialism, so that they understand our plight. This was a major paradigm shift in my world view. I was to run into another Irish Catholic priest exactly fourteen years later. This Irish priest Rev. Fr. Kevin O'Hara is a human rights activist and he was doing human rights from a unique perspective. In fact he saw human rights abuses as conflicts that needed to be resolved. I joined him and he changed my world. .

In 2001 I visited Ireland at the invitation of the World Jurists Association. I delivered a paper at that conference titled: *human rights, the judiciary and the rule of law in an emerging democracy*. My second visit to Ireland was in 2006. Again it was to attend a conference and to deliver a paper. I did more than deliver a paper. I saw as I put it at the community meeting in Erris "the emperor in his full nakedness".

When I visited Erris and Terry Conway took me on a tour of the pipeline route, I knew that I was part of history. I was convinced that what I have seen must be shared with people of the world. This book therefore is a descriptive analysis of what I saw in Erris. Initially I wanted to develop a manual from the activities of the people of Erris and use it to train the communities in the Niger Delta. Later I found that a manual may not be able to capture some of the more intricate nuances of what I saw, heard and felt in Ireland. I decided that a booklet of this nature will be a good starting point. I hope that you will agree with me after reading it.

I am doing this book with a view to show the inter-relatedness of the human experience. Secondly, this book is my own attempt to once and for all to relegate to the dustbin of history all the remaining relics of a conspiracy theory against the black race. If it is colonialism Ireland has seen it. If it is the division of their nation Ireland has seen it, in fact that of Ireland is unique because while colonialism brought Nigerians together, it split the Irish people. Ireland has seen poverty, economic and forced emigration and major epidemics. If it is multinational oil company perversions, Ireland has seen it. The lesson for us in Nigeria is that in spite of all these and with very little population, Ireland has remained one of the most prosperous nations in the world. What is the trick of Ireland in surmounting daunting situations? This book is my contribution to answering this question.

I am from an oil-bearing community. Chevron has been in my community since 1990. But that is not why I am involved. I am involved because what it takes to do community work is to get angry at injustice. And when I see injustice I recognize one. Moreover, I have acquired the requisite skills that have empowered me to understand that when I see something, I could do something about it. I have also been involved in issues in and around the Niger Delta for more than twelve years now. I bring that experience to bear on this work.

The Niger Delta of Nigeria shares a lot in common with the people of Erris. I hope to explore the similarities and differences of both experiences. I hope that by doing this the people of the world may be persuaded to come up with mandatory frameworks for monitoring and regulating the behaviour of multinational corporations especially extractive industries. I also hope that multinational corporations should also be able to examine their behaviours towards their host communities with a view to ensuring peaceful coexistence.

Finally theorists and practitioners of corporate social responsibility have looked at it from when companies are settled in a community and already doing business. To my mind this should not be so. We must segment the discussion into: entry behaviour of corporations, their behaviour when they have settled in, their exit strategies when they are leaving and how they should address legacy issues. And Erris and the Niger Delta provide clear examples of the implications of ignoring this segmentation in the discussion.

Methodology

The main methodology for collecting materials for this work is through interviewing (Kvale 1996). I interviewed about twelve people. Many of the interviewees were people from Erris. The interviews were not structured or recorded. They were more of extended conversations that lasted between 2 hours and 30 minutes. Most of my subjects were people involved in the *Shell-to-Sea Campaign*. I made notes of most of these conversations. In the Niger Delta I have interviewed people literarily for ever. In fact I have written a book on the Niger Delta conflicts (see Onuoha 2005).

I also participated in and observed some of the community's activities (Marshall & Rossman, 1999, pp. 106 & 107). For instance, I spent almost an entire day at the *Trailer*. I also attended the meeting of the forum for one evening. I undertook site visits. I spent close to three hours inspecting the pipeline route. I think that the most important and profound experience for me was visiting Erris to see and feel what they have experienced.

I have lived in the Niger Delta for more than forty years and worked in the area for more than twelve years. I have participated in different activities of the people of the Niger Delta. I have written four major works on the Niger Delta. I have experienced the Niger Delta both as an activist, academic researcher, peacebuilder and independent development consultant. I have drawn on all these to carry out this comparative study.

To my mind there are two basic models of analyzing conflict. First, is to categorize conflict into four different parts namely; pre-conflict situation, causes of the conflict and the conflict itself, conflict resolution stage and finally the post-conflict stage. Of recent the UN has added another stage – the peace consolidation stage. Pre-conflict stage deals with understanding the situation prevalent before the conflict erupted. This will help us to understand what changes led to the conflict (Linstroth, 2002, pp. 159-167). The causes of the conflict and the conflict situation deals with what we could call the root causes and the triggers of the conflict. Analyzing this also gives us an insight into the nature, dynamics and possible peacebuilding measures after the resolution. The resolution stage deals with what interventions were used by the parties to try and resolve the conflict. Finally, the post-conflict stage deals with the rebuilding of relationships and infrastructure damaged as a result of the conflict (Junne & Verkoren [eds.] 2005). Lederach (2005, p.43) has dismissed this post-conflict stage as an oxymoron. That should not detain us in this study. Peace consolidation is a relatively new concept in the field of peace and conflict studies. It is also different from post-conflict reconstruction and peacebuilding. Peace consolidation deals with how to "hold and keep the peace that has been achieved."

Another model of analyzing conflicts is using the simple category of location/context, parties, issues, dynamics, nature/structure, resolutions and post-conflict reconstruction. In analyzing and comparing the two situations I shall use both models to try and explain what the interviewees were saying to me in the interview.

I undertook a content analysis of several documents (Druckman 2005, pp.257-260). Some of the documents I got from the community and others I got from sundry sources that will remain anonymous for now for reasons of confidentiality and because the substantive case is still in court. This dilemma applies both to the Niger Delta and Erris.

My analysis shall be historical, descriptive, narrative and comparative. The main purpose is to create awareness of what is happening in both contexts and to see what each group could learn from the other. And most importantly to create a web of networks and collaboration that will reinforce our shared humanity.

Ethical issues

This book is not balanced in terms of data collection because I collected my data from mainly those that oppose the project. Two reasons account for this. First, I did not have access to the Pro-Erris Gas Group because of limited time and funding. Second, multinational corporations are usually hierarchical and closed. It is always difficult to get information from them. Even when one tries, they give you refrigerated answers or refer you to one staid publication or their web site. Instead of waiting to be referred, I visited these web sites myself and picked what helped my analysis.

I have been involved in issues around extractive industries, corporate social responsibility, human rights and peacebuilding for more than a decade now. I am not a fan of oil companies. This reflects in this book. This is not saying that this book is a propaganda sheet, not in the least. What I have done is to abide by the finest traditions of scholarly research and analysis by letting my sources speak for themselves with little interference from me.

Most of my informants shall remain anonymous for two reasons; first I did not seek their consent to use their names. Second, this matter especially as it concerns Erris is still in court so I do not wish to be charged for contempt of court. On the other hand there is so much going on in the Niger Delta. The security situation is also a cause for worry. I have a responsibility to protect those who provide access and information for me to do my work. However, some information on the Niger Delta are already public, if that is the case with an issue, confidentiality issue will not arise. Another unique thing about the Niger Delta is that even when you maintain confidentiality, by merely describing the case or issue, people might easily recognise the source and case in question.

Finally, over the years I have grown more matured both in age, experience and in my views about issues. I no longer engage in the blame game. I have seen good human beings both in government, oil companies and the communities. I have also seen the worst of creations in communities, governments and multinational corporations. Even among us, the civil society people who have come to see themselves as messiahs driven by our sense of self-righteous, there still so much to be done, to learn and to share. The most inspiring thing for me on this project is that are still so much to be learned.

Chapter 1

Introduction

At the World Economic Forum in 1997, the UN Secretary-General Kofi Annan declared, "in today's world, the private sector is the dominant engine of growth, the principal owner of value and managerial resources. If the private sector does not deliver economic growth and economic opportunity equitably and sustainably around the world, the peace will remain fragile, and social justice a distant dream. This is why I call today for a new partnership amongst governments, the private sector, and the international community" (quoted in Traub-Merz and Yates {eds.}, 2004, p. 152).

The challenge with the above declaration is how corporations can draw a line on what their obligations are since nation states seem to have the primary responsibility of promoting and protecting human rights and doing development. But a more discerning reading of the UN declaration has put the responsibility of protecting and promoting human rights on "every individual and every organ of society" (quoted in Business Leaders initiative on Human Rights, Report 2, Work in Progress, p. 7).

Theories of conflict have evolved over the years. In 1957 when the Journal of Conflict Resolution started, the main preoccupation was interstate conflicts. With the intensification of the Cold War, emphasis shifted to peace research which is basically concerned with the prevention of nuclear war. With the collapse of the Berlin Wall in 1989, studies of conflict focused on internal or civil wars (Crocker, Hampson and Aall {eds.} (1996).

A major aspect which has been missing is the conflict between groups or institutions in a nation-state where the state is not directly involved but is a bystander (Staub, 2003)[1]. And the state finds its hands tied because of several factors. A good example of this dilemma is Nigeria. The conflict in the Niger Delta or oil producing regions of Nigeria is mainly between the oil companies and the communities. The government needs the oil companies for revenue and investment; it also needs the people of the Niger Delta for their resources and for population for strategic reasons. The oil companies have international linkages which mean that Nigeria must tread with caution in dealing with them. And the people of the Niger Delta have also appealed to such universal values as human, environmental, ethnic and minority rights to justify their agitation. Caught in this web there is a need to reconceptualize these 'new conflicts'.

This reconceptualization is important as Ross (1993, p.96) notes, "actions of others, create real threats which promote in-group solidarity and collective responses against opponents." Conflicts and natural resources have attracted the attention of scholars and policy makers (Ross, 2001, Collier & Hoeffler, 2001, BICC, Brief 32, 2006). Karl (1997) spearheaded this debate when she argued that the source of conflict in resource rich nations is as a result of the 'paradox of plenty'. Gary and Karl (2003) reinforced this view. These two publications though influential in and around extractive industry circles, largely ignored the relationship between oil companies and their host communities. They rather focused on government's lack of capacity to transparently manage oil revenues. Desilier emphasised this issue of lack of capacity in her article titled: *Capacity Building and Oil Exploitation in the Gulf of Guinea* (Traub-Merz and Yates {eds.}, 2004, pp. 189-202). In that article, Desilier has supported the view that most developing countries that have oil resources do not have the capacity to (a) manage

---

[1] For a detailed discussion on the "concept of bystander" see Staub, E. (2003). The Psychology of Good and Evil: Why Children, Adults, and Groups Help and Harm Others. Cambridge, U.K. and New York: Cambridge University Press.

an oil economy (b) negotiate favourable agreements with oil multinational corporations (c) manage sudden huge oil revenues (d) develop and diversify the economy against price shocks.

This may also have influenced the publication by CRS and the World Bank (2005).which made the point that "the Chad-Cameroon Petroleum Development and Pipeline Project...... represents the foremost test case of the extent to which oil revenues can be used to alleviate poverty in a difficult developing country context" (p.4). The Managing Director of Shell Nigeria, Basil Ominyi shares this view when he declared in Shell 2004 Annual report that "in my view, poverty remains the over-riding problem in the Niger Delta and throughout Nigeria" (p.1).

On the other hand, Woolfson & Beck (2005) argue that the conflicts between oil companies and the host communities is as a result of the failure of corporate social responsibility in the oil industry especially in the area of health, work and the environment. Again the flaw in this work is that it does not examine the framework on which the issue of corporate social responsibility should revolve. Woolfson and Beck look at corporate social responsibility as a given without defining viable theories for the obligations of businesses in this context. Part of the obligations of oil companies is to protect the environment where they operate. Judge Weeramantry of the International Court of Justice at The Hague surmises that, "the protection of the environment is ...a vital part of contemporary human rights doctrine, for it is a *sine qua non* for numerous human rights such as the right to health and the right to life itself. It is scarcely necessary to elaborate on this, as damage to the environment can impair and undermine all the human rights spoken of in the Universal Declaration and other human rights instruments" (ICJ, Report 7, p. 4, September 25, 1997)[2].

---

[2] See also the judgement of Justice V.C. Nwokorie of Federal High Court, Benin City, Nigeria of November 14th, 2005 on the case of Jonah Gbemre (for himself and on behalf of Iwherekan Community) Vs.

This is why oil companies must develop an environmental impact assessment report before commencing operations.

From a legal perspective, Okonma (1997) argues that "statutory regulations of oil industry activities in Nigeria provide little or no protection for victims of oil pollution" (p.2). In support of this view Adewale (1989) insists that "victims of oil spill have claimed that the compensation paid to them is unreasonable and cannot be said to recompense for their loss. Oil companies on the other hand claim that the compensation paid to the victims is adequate........" (p. 2).

There are several statutory provisions for oil pollution victims in Nigeria (Onuoha, 2005, pp.139-144). The issue at stake as Frynas (2000) points out is the capacity and willingness of governments to ensure justice for victims. This unwillingness or inability as the case may be is why communities have sought extra-judicial means to protect themselves. One of these mechanisms is the memorandum of understanding (MOU).

Extending the argument further, Frynas (2000) insists that a major source of conflict between the oil companies and communities is that the communities cannot get justice whenever they resort to litigation in their disputes with oil companies. In support of Frynas, Okoli (1994) argues that "it is a self-evident fact that the attitude of organs of the state to the prevention and prosecution of white-collar offences are generally lukewarm" (p.2). Though Frynas studied 116 court cases between oil companies, and communities, it makes his argument very persuasive but not convincing. It is not convincing because the Niger Delta communities are not 'litigatious'. Second, they are poor and do not have the resources to pursue protracted legal battles. And as Frynas notes, the people do not have confidence in the legal system.

SPDC, NNPC and the Attorney-general of Nigeria (The Guardian, Tuesday, April 25, 2006, p.47.

Okonta and Douglas (2001) and Welch (1995) have situated the Niger Delta issue within a human rights discourse. While Welch makes the case for self-determination using the Ogoni as case study, Okonta and Douglas situates the discourse around human rights violation perpetrated by the Anglo-Dutch multinational Shell. Osaghae (1991, 1995 and 1998) pursues the same argument. There is no doubt that there are human rights violations in the Niger Delta. But the question is why is it that the people of Aceh in Indonesia for instance had the desire and even fought for self-determination but never targeted the oil companies while the people of the Niger Delta vent their spleen on the oil companies? This is not to suggest that communities must respond to issues in the same manner, however, if the issue in the Niger Delta was basically that of human rights violations expressed in the form of self-determination, the culprit to hold responsible is not the oil companies but the government.

A more interesting dilemma is why is it that the people of Erris have never muted the idea of self-determination but rather have insisted on Shell not coming in at all? Does it mean that to the people of Erris their membership or citizenship of the Republic of Ireland is a given or closed issue. This is important because it may be extremely difficult if not downright impossible to dismiss the issue of the relationship between oil companies and host communities from the lens of governance.

Other scholars (Atsegbua, 1993 & Abusharaf, 1999) have looked at the legal agreements binding nation states and oil companies and concluded that since these relationships are unequal that they could explain conflicts in the oil-producing regions. While Abusharaf makes a historical analysis using the Sudan as case study arguing that "the resource sectors of developing oil-producing countries were tied up by a series of unequal contracts governed by the dominating traditional concession regime" (p.4), Atsegbua undertakes a legal excursion using the various agreements between Nigeria and multinational corporations as his case study and concluded that "the Production Sharing Contract certainly has no benefits whatsoever to the NNPC as it stands today....." (p.10). The cardinal question which remains to be resolved is that the conflicts are not between the nation states and oil companies but between oil companies and their host communities. But in the case of Ireland, the people of Erris claim that the state is not participating at all not to talk of agreements that are not beneficial.

Two scholars Murshed (2002) and Azam (2001) have developed penetrating insights into the dynamics of conflicts which to my mind will assist me in analyzing the conflicts between oil companies and communities in the Niger Delta and Ireland. Murshed argues that "fundamentally, violent conflict is a symptom of the absence or breakdown of the implicit or explicit social contract that sustains peace" (p.3), Azam concurs with the following "violent conflicts must be thought of as resulting from a failure of the state to perform some of its fundamental tasks...." (p.3). As Imobighe notes the communities "blame the government for failing to protect them and their environment from the adverse consequences of oil exploration. The people believe that there is a conspiracy between the government and the oil companies to get oil out of the region at all cost irrespective of what happens to the people" (Traub-Merz and Yates {eds.}, 2004 p.106).

In conclusion the debate revolves around whether there should be voluntary or mandatory regulatory frameworks for multinational corporations. The BICC (2006, p.4) quotes Billon (2003, p.230) as suggesting what they refer to as co-regulation, that is mixing voluntary membership with mandatory compliance. The point that is to be made from the short review above is that whether it is self-regulation, co-regulation or mandatory regulation, the most critical element in managing oil company and community relation is the human factor. As we shall see, this piece has been missing This is despite the fact that in March 2001 NBC News quoted a *Wall Street Journal* poll which claimed that 62% of Americans were opposed to drilling in the North Slope. The report concluded that "the people who run the *Filthy Four*[3] have repeatedly placed their selfish interests above all other considerations, not only within the US but around the world" (Corporate Campaign Inc. 2001, p.1). So the question is which one is more important the preservation of our collective humanity including the environment or mega profits for a few?

Brief Background of Shell's Presence in Erris, Ireland

The Republic of Ireland is a small country located in the extreme north-west of Europe. The capital is Dublin. It is on the west of Britain. It was colonized by Britain for more than 300 years. It eventually gained independence on December 6; 1921.There is the Northern Ireland in the north and the Republic of Ireland in the south. Erris is located in the Republic of Ireland. The country has a total land mass of 68,890 sq. km. It is endowed with several natural resources which include gas, peat, copper, lead, zinc, silver, barite, gypsum, limestone and dolomite.

The Republic of Ireland has a population of about 4 million. Infant mortality rate is 5 deaths per 1000 live births. Total life expectancy for both male and female is about 77 years. HIV/AIDS prevalence rate is 0.1% as at 2001. In the same year the country had about 2,800 people living with HIV/AIDS. 98% of the total population is estimated to be literate, while 88.4% of the population are Christians of Catholic extraction. English and Celtic are the two main official languages.

The economy is strong and experienced 7% growth in 2004. Industry accounts for about 46% of GDP, 80% of exports and 29% of labour force. 5% of GDP is from agriculture, 46% from industry and 49% is from services. Republic of Ireland has average budget revenue of $70.46 billion and expenditure of

---

[3] Filthy Four refers to the four oil majors namely Shell, Chevron, ExxonMobil and Texaco.

$69.4 billion. From the above and considering the population of Ireland, it is a rich country. I make this point because it has been argued and even emphasised that poverty is the main cause of restiveness in the Niger Delta (see Shell Annual Report 2005). This also explains why development and "poverty reduction" have become the main strategy for responding to conflicts in the Niger Delta.

The government is democratic with bicameral legislature. Republic of Ireland has four provinces namely Connacht, Leinster, Munster and Ulster. It is divided into 32 counties. Six of these counties are still under British rule in Northern Ireland. Carlow, Cavan, Clare, Cork, Donegal, Dublin, Galway, Kerry, Kildare, Kilkenny, Laois, Leitrim, Limerick, Longford, Louth, Mayo, Meath, Monaghan, Offaly, Roscommon, Sligo, Tipperary, Waterford, Westmeath, Wexford, and Wicklow. It has a ceremonial president and a prime minister.

Many scholars have argued that democracy is a major instrument for peacebuilding and that democratic nations are largely stable. I agree but as we shall see in the case of Ireland, what democracy did, was to minimise if not totally eliminate the use of violence as response to the conflicts in the area. Many have also attributed the protracted nature of the conflicts in the Niger Delta to the lack of democratic structures and culture. To what extent this could be true shall be examined in this book.

Erris is in Mayo County in the west of Ireland. The capital is Castlebar. Mayo County covers a land mass of about 5,397 sq. km. and has a population of 117,428. Erris is a rural riverine community. The area has suffered neglect for centuries. This led to massive emigration from the area in the second half of the 19th century. The people of Erris are mainly farmers and fishermen/women. This is how the Centre for Public Inquiry in its report on the Corrib Gas Project describes Rossport – which is a small town in Erris where the controversy took its name from; "Rossport is in the parish of Kilcommon, which lies between Belmullet and Ballycastle along the north coast road in County Mayo. Along the coast, the sea has carved spectacular cliffs that rise to more than 300 metres. The Blue Stack Mountains rise to the north, and Benbullen and Knocknarea rise to the east. This region, with its megalithic tombs and stone circles, is one of the oldest inhabited areas of the world. The world heritage site at the Ceide Fields along the coast road celebrates a prehistoric landscape that contains the world's oldest known field systems, the early agricultural endeavours of five thousand years ago. It is an unpolluted, sensitive and scenic landscape. The air is clean and much of the local population draws its drinking water from Carrowmore Lake. Several locations around Sruwaddacon Bay and the planned processing plant site are designated or proposed protection areas. The proposed landfall site for the pipeline at Glengad beach at Broadhaven Bay is a proposed candidate Special Area of Conservation (SAC) and a designated Area of Special Scenic Importance. The Glenamoy Bog Complex including Sruwaddacon Bay is an SAC, as is Carrowmore Lake. Pollatomish Bog is an SAC and a proposed Natural Heritage Area" (p.26).

In January 1993 Enterprise Oil a British firm and two Norwegian firms were granted a licence to prospect for oil and gas in the Slyne Basin. Prior to this time the Department of Energy for over two decades has collected seismic and drilling data about this area. This information according to sources was sold to this new consortium led by Enterprise Oil for the sum of £8000.

In the introduction, I have argued that we must look at corporate responsibility proactively, instead of looking at it when corporations have already settled down to business. It is obvious that they people of Erris were never part of this licensing round. They had no say in it. This is at the base of the misunderstanding between communities and corporations. If communities are involved from the start, it could go a long way in minimising the areas of friction and misunderstanding between them and corporations. The question still remains what is the place of communities in all this?

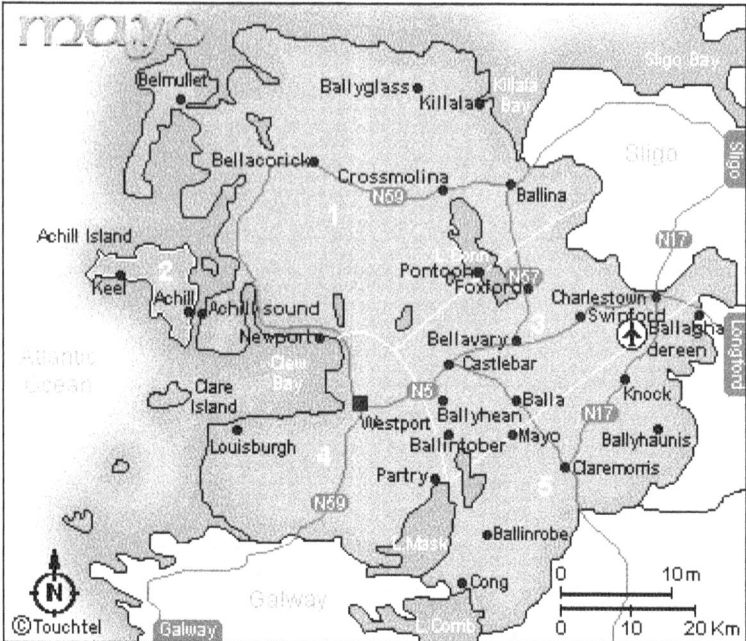

Map of County Mayo courtesy of Mayo County of Ireland.

The Corrib Gas Field is equivalent to about 182 million barrels of crude oil. Its worth is estimated at €50.4 billion. As the CPI report put it "the market value per one trillion cubic feet of Corrib gas is €8.4 billion, putting the potential value of the Corrib and surrounding fields for Shell and its partners in excess of €50.4 billion" (p.23). Equity participation in the project is broken down as follows: – Norway (Statoil) 36.5%, Shell 45%, Marathon 18.5%, Ireland 0%").

In 1996 the consortium announced that they have found gas in the Corrib Field in the Slyne Basin. During this three years of prospecting, apparently the community was never part of the whole arrangement. In 1999 the consortium requested to purchase a piece of land at Ballinaboy, in Mayo County to build a processing plant and for a pipeline to transport the untreated gas. In 2000 the consortium approached the State Forestry Service to acquire a 400-acre site at Balliinaboy. Before this time the Gas Act allowed only the Bord Gais to construct a pipeline. This Gas Act was amended in 2000 to permit the consortium to proceed with project. In the same 2000 the regulatory power for the oil and gas industry was transferred from the Minister for Public Enterprise to the Minister for the Marine and Natural resources. In 2001, the consortium applied for a petroleum lease to commence operations. In August 2001, Mayo County council granted a planning permit to the consortium to construct an €800million gas processing plant on the 400-acre land. According to the plan the processing plant was to occupy only 23 acres.

In November the new minister introduced another statutory instrument which gave him power to make Compulsory Acquisition Orders (CAO). According to the Centre for Public Inquiry,

"this meant that the CAOs made by the minister permitted a private company to occupy land and construct a pipeline even if the owners of the land objected. Within weeks, landowners along the route of the proposed pipeline were informed that they would be served with CAOs unless they accepted compensation and allowed EEI to lay the pipeline" (p.14).

In March 2002 this new order became part of the Gas Act. And it was in the midst of this emerging confusion that Shell acquired Enterprise Oil for the sum of €6.5 billion. A month later the minister granted permission for the construction of the pipeline.

When this was made public, residents of Rossport filed objections opposing the construction of the pipeline and the gas processing plant. However the most scathing indictment of the project came from Kevin Moore a senior planning inspector. In his words, "from a strategic planning perspective this is the wrong site. From the perspective of government policy which seeks to foster regional development, this is the wrong site, from the perspective of minimising environmental impact, this is the wrong site, and consequently, from the perspective of sustainable development, this is the wrong site".

Moore concluded his report by insisting that the Marine Licence Vetting Committee set up to examine the environmental aspect of the project has failed to address the demands of the objectors. Mr. Andy Pyle the Managing Director of Shell Ireland in his own defence said that it was not economically viable to build the processing plant on a shallow water platform and that it would escalate the cost to about €360 million.

This list of grievances was to expand as events unfolded. In April 2003 the board had another hearing and it was here that they rejected the planning permission that was granted because "the transfer of 600,000 cubic metres of peat bog to land near the proposed processing plant would represent an unacceptable risk and could pollute local rivers". With the withdrawal of the planning permission, Shell complained that the delay in granting a fresh permit has escalated the cost of the project by €100 million.

While Shell and other members of the consortium were trying to lobby the authorities to get a new permit on September 19, 2003 "a massive landslide drove tonnes of peat and mud off Barnacuille and Dooncarton mountains overlooking the route of the pipeline, sweeping away homes and the remains of the deceased in a nearby graveyard. The landslide covered one of the original proposed pipeline routes which, if implemented, would have been carrying unprocessed gas were it not for the planning delays' (p.16).

Pushing on still with the project, in December 2003 the consortium submitted a new planning application to the Mayo County Council. This time they proposed to move the peat to

Bord na mona. In April of 2004 the planning permission was granted. Again the residents continued their protest.

In April of 2005 Shell went to court to seek for an injunction against residents who were obstructing the laying of the pipeline across their lands. On June 29, 2005 Micheal Seighin, Willie Cirduff, Vincent McGrath, Philip McGrath and Brendan Philbin were sent to Clover Hill jail in Dublin for failing to comply with an order of the High Court restraining them from interfering with Shell & P's attempts to lay the pipeline' (p.17).

With the jailing of the *Rossport Five* the conflict has turned full circle. The residents now formed the *Shell-to-Sea Campaign*. This was targeted at the following:

1. That the pipeline was too close to their homes and farms
2. That it was risky since the pipeline will carry untreated gas
3. Threat to the sensitive and scenic location
4. The instability of the peat bog
5. The risk of a major accident

I shall now try to address some of the issues raised using materials from the story and data from the interviews and my visit. In doing that I shall be comparing the situation in Ireland with that of Nigeria. Before then, a few observations will help to properly situate the analysis. First is that what is happening in Erris is only in a small part of Ireland while in Nigeria, the entire Niger Delta is affected by various oil company activities working as a consortium. Moreover, in Nigeria, there is oil drilling and exploration, gas gathering and processing plants and even crude oil refining plants. Increasingly business activities related to the oil industry in Nigeria is becoming very prominent and overshadowing other activities. Before that let us do an overview of Nigeria. The essence of this overview, is to also situate the discourse within the Nigerian context. This overview will also provide the background information on the oil industry in Nigeria especially for those that are not very familiar with the issues.

Oil in Nigeria: An Overview

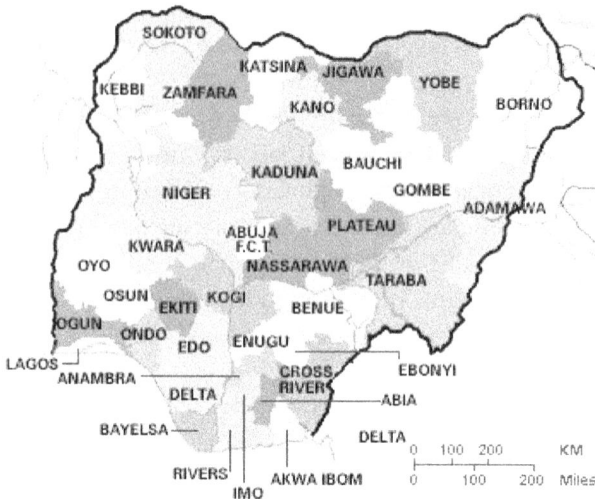

Map of Nigeria showing the 36 states. Courtesy of Shell Nigeria.

Nigeria is the fifth largest oil producing nation in the world. It has about 20 billion barrels of crude reserve (CRS, 2003, p.26). Oil was found in 1956 at Oloibiri in the present day Niger Delta region of Nigeria. Nigeria's oil is explored and exploited by multinational corporations. However, the activities of oil companies are supposedly regulated by the Nigerian National Petroleum Corporation (NNPC) which is a government parastatal. In practice this is not so. The paradox of the Nigerian situation is that the oil industry regulates itself. Most times the government regulatory agencies plead lack of fund, equipment and personnel to carry out their oversight functions. However, the debate among practitioners and academics has been whether government is unable or unwilling to regulate the oil industry. Elsewhere, people have argued that the Nigerian government and its agencies lack the capacity to regulate and monitor oil industry activities. But the oil industry has the reputation also of being the most regulated

30

sector all over the world. We shall see the implications of these as the discourse unfolds.

This is how Omeje (2006, p. 478) puts it, "even though oil is most strategic to Nigeria's economy and politics, it is significant that the Nigerian state has been unable to evolve a coherent and comprehensive policy framework for the management of oil resource and their negative externalities, the distribution of the accruing revenues or for engaging with the increasingly restive oil-bearing communities of the Niger Delta." I do not intend to join issues on this matter since it is not the main focus of this book. I raised this issue so that we can bear it in mind as an important conversation which we must engage in as we reflect on this oily situation. However, the interesting thing here is that the Republic of Ireland evolved many legislations and policy framework to regulate the oil industry but still it did not prevent the communities and oil company from misunderstanding each other. The issue here to a large extent is not the absence of laws and regulatory frameworks but the capacity to enforce, the willingness to regulate and the political will and sagacity to carry everyone along.

In 1960, Nigeria gained independence from Britain. Northern and Southern protectorates of Nigeria were amalgamated in 1914 by Lord Frederick Lugard the colonial governor. Nigeria has about 250 ethnic groups with many language variations. After independence, in order to become a modern state, Nigeria became a federation. This was used to integrate all the ethnic groups into a modern nation. Whether this strategy employed by Nigerian political elite has worked or not is there for all to see. From three regions, in 1963, the fourth was added, Nigeria today has 36 states. Apart from the 36 states, Nigeria has 774 local government areas. Out of 468 years of independence, the military has ruled Nigeria for 29 years (Falola, 1999, pp. 1-16). Still the agitations for more states seem not to be abating.

The 2006 census puts Nigeria population at about 126 million. Out of Nigeria's population of about 140 million, the people of the Niger Delta constitute about 20 million. The Niger Delta is located in the south eastern part of Nigeria along the Atlantic coastal plains. The main occupation of the people is farming and fishing. In my other book, I have argued that one of the causes of conflict in the region is that oil exploration activities have disrupted their sources of livelihood without providing an alternative (Imobighe, 2002). Second, I had also insisted they want to control the resource in their area or put in other words, they want to have a say in how the revenue from oil is distributed. This demand is not without justification. They argued, albeit, correctly that when the other regions had their palm oil, cocoa and groundnuts, they controlled them (Eghosa, 1999).

Another major complaint of people living in the oil-bearing communities is that of pollution and environmental degradation. In Erris of Ireland this was the main issue why the people opposed the Corrib Gas Project. The threat of environmental disaster is real. But for some reasons, the authorities seem too eager to minimise the issue. But for people from the oil-producing regions it is a matter of life and death. From Erris to the Niger Delta environmental issues are some of the main causes of the conflicts which we are experiencing.

Responses to these issues have been varied (Onuoha, 2005). For the government, law and order must be maintained[4]. So as far as the various governments are concerned, the people of the Niger Delta are breaking the law. So government's response has been to brutally suppress any agitations. Community responses have evolved over the years. It ranged from writing letters to taking oil company staff hostage, from the barricade of oil installations to mass uprising (Wokoma, undated).

---

[4] See also Report of the Special Security Committee on Oil Producing Areas dated Feb. 19th, 2002. See also Background Papers for Niger Delta Peace and Security Strategy published by Peace and Security Secretariat, Port Harcourt, Nigeria, December 2005.

32

Kidnappings, hostage-taking and assassination of oil workers continue unabated. Sabotage of oil installations and facilities are still prevalent. On Thursday May 11, 2006 some AGIP oil workers were kidnapped in Port Harcourt. On Wednesday May 10, 2006 an American working for an oil servicing company was assassinated in Port Harcourt. To confirm that the kidnappings and assassinations were not only targeted at westerners, on Tuesday June 20, 2006 two Filipino oil workers were kidnapped and subsequently released on Sunday June 25 (see Vanguard, Monday, June 26, 2006, p.15).

A new dimension was recently added to conflicts in the Niger Delta. In May 2006, there were two bomb blasts in Port Harcourt and Warri (see Daily Independent, Monday, May 8, 2006, p.A13). What was worrisome was the fact that the explosion in Port Harcourt happened in a military cantonment. The implications of these are that there is a possibility that ex-military personnel were involved. In fact the main access road that runs through the barracks has ever since been closed to the public. But the question still remains as to who was responsible for the blasts?

What is even more worrisome is that most of these new tactics of responses is being copied from places like Sudan, Iraq, Colombia and Nicaragua. The implications of these is that while the government feels that nothing much is happening, the militants have already convinced themselves in their mindset that they are waging a war. The government on its own part cannot go for an all out war because the enemy is not immediately visible. And there are human rights and other implications of such heavy handed response.

Oil companies' responses have been characterised by short-sightedness and are more often than not perfunctory. They have also used the law as back-up for their responses. Oil companies are also often accused of impunity and a total disregard for the law. Oil companies have also evolved different strategies for responding to conflicts in their areas of operation. For instance, Shell Nigeria claims that "similar

33

challenges have also informed paradigm shifts in strategy over time from a traditional philanthropic model (community assistance), to community development (strategic social investment) and more recently, sustainable community development" (Traub-Merz and Yates [eds.], 2004, p.143).

Conflicts between oil companies and their host communities have been on the increase especially in Nigeria (see Azaiki, 2003, Imobighe et al, 2002, CDHR, 2000, Okonta & Douglas, 2001). This has led to hostage-taking, vandalisation of oil exploration facilities and increase in crude oil prices. The main reason offered by the communities for the restiveness in the area is "the exploitation of these resources is not in the hands of the people of the region. The Niger Delta people strongly believe that the Nigeria state has failed to represent their interest in the manner the prosperity generated from their place is being distributed and as such would want to take over the control of these resources themselves" (T.A. Imobighe in Traub-Merz and Yates [eds.], 2004, pp. 101 & 102). This very issue of resource control is core to understanding the conflicts in the Niger Delta.

One aspect of the conflict in oil producing regions which seem to have eluded analysts is the economic dimension of the conflict. In 2006, according Edmund Daukoru Nigeria's minister of state for, Nigeria was losing 850,000 barrels per day due to the conflict. This translated to close to $300 million per day. Against the backdrop of the Nigeria economy which is almost 90% dependent on oil, the impact of this is better imagined. Again Nigeria's democracy is evolving and people's expectations of the democratic project seem to have waned. With this significant drop in oil revenue and the cost of running government in Nigeria and the problem of corruption, it is obvious that soon people will question the ability of democracy to meet their needs. This could lead to mass uprising and a resentment against politicians and the self-fulfilling prophesy of the CIA that Nigeria as a nation may not last for the next 16 years.

Map of the Niger Delta showing the different ethnic groups: Courtesy of Shell.

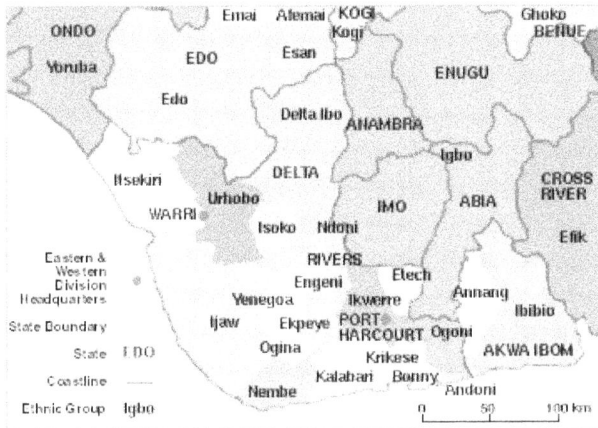

Chapter 2

Comparative analysis

Pre-conflict stage

Using the two models proposed above let me now analyze the conflict. I will first look at the pre-conflict stage or the location/context of the conflict. Erris is a beautiful and serene community off the coast of the Atlantic. It is like three tea saucers lying on each other with the Atlantic Ocean surrounding it. There are undulating hills that dot the landscape. It is a rural community where everyone knows each other.

The people of Erris are mainly farmers and fishermen and women. The air is clean and the environment inviting. This is where Shell found gas off the Atlantic Coast. It is in this community that Shell wants to build a pipeline and a gas processing plant. Most of the people of Erris are native Irish speakers and were born in the area. However, a few people from other European countries have migrated to and settled in Erris. For instance, I spoke to a man who is originally from Germany. Many Irish-Americans have also decided to come back and settle in Erris.

From the above description, this community is not ready and was not prepared for the changes that will inevitably accompany oil exploration. It is also obvious that the people feel that they were not ready for strangers yet. The anticipated population explosion of the area is a prospect that the people are not ready for. As McAdams, Josselson and Lieblich (eds.) (2002, p. xv) noted " ....how ill-prepared we may be for change, how we welcome change and how we dread it". This is how my respondent put it "I come from here, look out of that window, if Shell comes in here they will destroy everything, we know their history" (interview with Erris resident).

The Niger Delta on the other hand is among the three largest wetlands in the world. It is made up of the sandy coastal plain and mangrove forests. There are also the fresh water swamps that are either permanent or seasonal and finally there are the dry land forest areas. The Niger Delta has been described as one of the most fragile and sensitive ecosystems in the world. The Niger Delta connects the Atlantic Ocean through a web of creeks and rivers.

Why then is it that resources tend to 'hide' in these areas that are not easily accessible considering the similarities between the terrain of the Niger Delta and Erris? A possible explanation is that the world seems to have exhausted its resources especially those in accessible areas and are in search of hidden treasures in remote parts that have long been ignored.

However there is one big difference between the Niger Delta and Erris. While the government in Nigeria have denied the Niger Delta communities ownership of resources that are located offshore[5], Mayo County has been involved and is really in the forefront of granting planning permissions and enforcing all regulations guiding the operations of Shell in the area.

In analysing conflicts, it is not the physical location alone that is taken into consideration. Time and socio-political context are also important. Erris is a marginalised part of Ireland. This is how the CPI report describes Erris, "the barony of Erris has suffered centuries of neglect and mismanagement. In the second half of the 19th century, there was massive emigration from the area to Britain and the United States. The principal source of income in late 19th century Erris was money from relatives abroad, a flow of remittances which continued, like mass emigration, into the 1960s. Two thousand people live in Kilcommon, which contains one of the few Gaeltacht areas remaining in Ireland. The economy depends on small-scale

---

[5] For details of this controversy, the issues involved and the Supreme Court judgement visit www.nigeriafirst.org.

farming and seasonal fishing, and the area has consistently suffered from emigration because of lack of local jobs" (CPI Report, p.26).

This is how Okonta and Douglas (2001, pp. 33 & 34) described the Niger Delta,

"yet, in spite of its considerable natural resources, the area is one of the poorest and most underdeveloped parts of the country. The inhabitants, 70% of whom still live in a rural, subsistence characterised by a total absence of such basic facilities as electricity, pipe-borne water, hospitals, proper housing and motorable roads, are weighed down by debilitating poverty, malnutrition and diseases......annual income in the Niger Delta are still far below the national average. Historically, the people of the Niger Delta have always been at the mercy of greedy outsiders who plunder their natural resources without giving them anything in return, from the days of slavery to the present day."

As if to underscore and reiterate the above views, The Managing Director of Shell Nigeria, Basil Ominyi declared in Shell 2004 Annual report that "in my view, poverty remains the over-riding problem in the Niger Delta and throughout Nigeria" (p.1).

Many scholars have linked conflict to poverty. In fact World Bank (2004, p.14) notes that "countries affected by conflict face a two-way relationship between conflict and poverty – pervasive poverty makes societies more vulnerable to violent conflict, while conflict itself creates more poverty." My intention here is not to make a case of poverty as a source and cause of conflict, but demonstrate how two different communities with the same resources being managed by the same organisation could share the same fate. Second, comparing the poverty in Erris and the Niger Delta is also relative. While the people of Erris have gone past such basics as water, electricity and hospitals, Niger Delta communities can be described as people of the present living in the past.

Another area where the people of Erris and Niger Delta share similarities is in their history. Two observations are important here. First, is that the *Rossport Five* after the jail did no longer see themselves as mere residents of Erris. They reminded whoever cared to listen that imprisonment "historically has and will always fail". They linked the past with the future and this has great implications for the existence of the company in the area. Second, the imprisonment was used to remind the people of Ireland that Shell is not indigenous and that Shell's relationship with the British even though Dutch is a reminder of more than 300 years of colonialism in Ireland. The Northern Ireland issue also reverberates. As one respondent told me, the government is being careful because whatever happens here may have great implications for the peace process in Northern Ireland which is still under British rule. By imprisoning the *Rossport Five* Shell unwittingly made martyrs out of regular citizens. And as they declared; "the campaign has now begun in earnest" (p. 49).

Anyone conversant with the history of the Niger Delta will recall that it has been a history of active resistance. As Onuoha (2005, p.77) puts it for the people of the Niger Delta

> "to have peace they fought wars. They went to war to capture slaves, and to prevent themselves from being captured and sold into slavery. They fought to be part of the so-called legitimate trade. They fought to preserve their communities from being run over. In short they fought for their lives."

It would also be recalled that in 1966 Isaac Adaka Boro declared the Republic of the Niger Delta. There is also the story of King Jaja of Opobo and how he resisted the colonialists.

A people with a strong history of resistance embedded in their identity will prove very difficult to engage in a conflict situation. Ross (1993, p.96) insists that

> "psychocultural conflict theory explains conflict in terms of psychological and cultural forces that frame beliefs about the

self, others and behaviour. These dispositions which constitute the inner worlds of a group and its members, are used to make sense of external events and the behaviour of others."

To what extent did the oil companies involved take into cognisance the historical realities of these communities? This is because a people who have suffered and waged wars to preserve their identity will have no qualms in treading the same part again especially when they have always seen themselves as having successfully resisted their oppressors. This historical overview is important as we move into the next phase of the comparison.

Parties to the conflicts

Identifying the parties to a conflict is part of understanding the structure of the conflict. There are several reasons why it is important to accurately identify the parties in a conflict. First, it gives insight into the kind of resources that are being mobilised for the conflict. Second, if the conflict is on-going as in Erris and the Niger Delta it shows the nature of the conflict. Parties to a conflict tell us whether the conflict is escalating or de-escalating (Kriesberg, 2003). Knowing the parties to a conflict also recommends possible resolution measures. Identifying the parties helps also to analyze the attitudes, characteristics and behaviours of the parties with a view to putting in place conflict resolution, transformation and peacebuilding mechanisms. The parties involved in a conflict also play an important role in resolving the conflict. If we understand the parties we will understand which party will play which role in ending the conflict.

In every conflict there are primary, secondary and tertiary parties. There are also shadow parties. All these play some distinct roles before, during and after the conflict. By their actions some parties that has come to help in resolving a conflict, unwittingly become parties in the conflict.

Shell Petroleum

Shell is arguably the largest multinational petroleum exploration and marketing company in the world. Shell is all over the globe. Shell is the main party in the conflict in Erris. Shell is the first company to find oil in Nigeria. Shell controls more that 40% of Nigeria's oil. Shell oil activities are on-shore. There are several implications of the role of Shell in conflicts. An interesting phenomenon is that the role of Shell in conflicts in the Niger Delta has evolved over time. As this role evolves so also is Shell's response to the conflicts in its area of operation.

In Erris, the people are looking at the legacy of Shell as an organisation. For instance, Shell is a foreign company. It is Dutch in origin and has a large presence in London. To the people of Erris, Shell represents a relic of their mournful colonial past which should be resisted at all cost. Second, the people recall the activities of Shell in places like Nigeria and Russia and vow that it will not be their portion. So from day one Shell was an unwanted guest in Ireland. Space may not allow an elaboration of this, but it is instructive that the people of Erris preferred to visit Norway to discuss their plight rather than go to Shell Centre in London.

In 1993 the people of Ogoniland in the Niger Delta declared Shell *persona non grata* in their area. In 1998, the Ijaws gave Shell a 48-hour ultimatum to leave their land. Other groups have extended such deadlines to Shell at different times or the other. All these can not be for nothing. There are several reasons why Shell has become an unwelcome guest in many areas and a major party in many conflicts.

The history of Shell, its colonial legacy, its structure, nature and culture have become burdens which make the corporation a perpetual party in conflict. And as they say wherever Shell has gone, its reputation has preceded it which is an albatross which it has carried over the years. In conflict situations Shell has had to bear the burden of this reputation. It has also made

Shell a target of community anger. For instance, since the new upsurge in the kidnapping of oil company workers in Nigeria, it is Shell workers and contractors that have been mainly victims. The reasons for these are not within the purview of this work, it is important to understand the central role of Shell in the conflicts in Erris and the Niger Delta.

From the above review, one thing that comes out clearly is that Shell has the influence and leverage to change its role in any conflict. It also has both the capacity and capability to play a more constructive role in conflict management not only in its area of operation but in our entire universe. The question which arises is why has Shell decided to have this image of a conflict-monger instead of that of a peacebuilder? Can individuals, organisations and nations decide the kinds of roles that they wish to play in conflicts? To put it blandly, can an organisation like Shell make deliberate decisions to 'avoid' conflicts?

The Communities

I deliberately did not mention the communities first as parties in the conflict because I think that if Shell has not entered the communities, may be the people would not have been in conflict with them. So it is Shell's activities and mobilisation in the communities that brought it into conflict with the people of Erris and the Niger Delta. As I mentioned in the introductory part of this work, conflicts between such groups and Shell and the communities have not received much attention in mainstream literature. To my mind the reason for this is that these conflicts (between communities and oil companies) have been subsumed under the general theme of intergroup conflicts. On the other hand, other scholars have interpreted it as conflicts between the nation state and its minority group. So it either qualifies as ethnic conflict or a quest for self-determination.

I think that these approaches do not capture the intricate and intrinsic dynamics of the conflicts between these communities and oil or other extractive industries. These conflicts pose a different kind of challenge. First, the nation states seem either unable or unwilling to do anything. Second, whereby they have tried to do something, it has been haphazard and aimed at placating the multinational corporations. Third, in the framework underlying the relationship between these companies and the governments, the communities seem almost non-existent. Finally, other components of the nation state seem unaffected by these conflicts or pay lip service of support to the communities at the receiving end.

The communities of Erris and Niger Delta share a lot in common. They are poor, rural and riverine communities. In a sense both communities are marginalised minorities. The communities consider their land the only standing resemblance of their identity which must be preserved at all cost. Both communities also feel neglected and do not trust the governments to protect their interests. This is not without justification for in the case of Ireland, the speed with which some of the planning permissions were granted left a sour taste in the mouth, while in the Niger Delta, the people have consistently been betrayed by subsequent governments starting from the colonialists to the military and now to civilians.

Finally, what makes the communities in Erris and Niger Delta unique as parties in these conflicts is the reinforcement of their identity by attachment to the land, rivers and other natural endowments in their environment. When parties to a conflict have their identity attached to their natural environment and one of the parties to the conflicts has an identity that threatens that environment, then the dynamics and nature of that conflict may be different from what we already are conversant with (Harris & Reilly {eds.}, 1998). This is the case in the Niger Delta and Erris. The people of both communities feel and sense a conspiracy to threaten if not

44

destroy their identity which is one of the few remaining relics of who they are having lost their dignity to prolonged neglect and marginalisation.

But what is a community and who make up communities? This question is important because in modern day business world, communities members are looked down on or are seen as plain illiterates, uneducated and 'unreasonable.' The fact however is that we all are community members and belong to different communities. It is interesting how our identities are constructed and re-constructed in conflict situations. In my work I have listened to employees of oil companies from the Niger Delta talk of the communities as if they are from the moon. And when community representatives manage to extract concessions from oil companies and governments, we all come home to enjoy the goodies.

In the Niger Delta communities are more of land and location specific made up of people who share ancestry, history, origin and culture. Communities are usually closed entities that do not easily admit new members. Even when new members are admitted they do not automatically get acceptance and enjoy privileges. In the Niger Delta communities also share language and religion.

But in Erris this is not always the case. Even though many Erris residents share common history and ancestry, there are also immigrants. For instance, as I mentioned earlier I saw a man from Germany who relocated to Erris and decided to make it his home. He participated actively in the *Shell-to-Sea* campaign. I did not spend too much time in Erris to get a clue as to his degree of acceptance by the group but from the little I saw, the man seemed an influential member of the group. The point which I am making is that we must not conceive communities as people who do not know what they want and thereby treat them disdainfully. This has been one of the sources of conflicts in the Niger Delta.

Governments

One of the main functions of government is to resolve conflicts between the various components units of the nation. One of the key characteristics of governments when they resolve conflicts is that of neutrality. In the fulfilment of this function, governments inadvertently become parties in such conflicts. In the case of Ireland and the Niger Delta governments were involved in the conflicts from the onset. By this I mean that governments did take sides during these conflicts. Apart from taking sides they openly supported oil companies. Let me situate this a little.

Oil business is big business and oil money is big money. Governments have always seen investment in oil as strategic national investments. This makes them to almost always over-pamper the oil companies. This reflects in the kind of concessions granted to oil companies. Whole national legislations have been changed to make space for oil companies. Many politicians see the coming of oil companies as the ultimate foreign direct investment.

In Erris there are two levels of government involved in the oil issue. First is the Republic of Ireland and the Mayo County. These two governments desperately wanted the oil investment. According to Bertie Ahern the Minister of Finance, some of these over generous concessions were "designed to improve Ireland's competitive position in attracting oil and gas exploration" (quoted in CPI Report, 2005, p.11).

In the Niger Delta, the situation is both similar and different. The Nigerian government has always been party to the conflict because; the law allows the government to control all mineral resources in the country. Most legal agreements or contracts concerning oil exploration and exploitation are usually between oil companies and national governments. In fact oil exploration activities are usually carried out between state enterprises and private multinational corporations. And in certain instances like Norway (Statoil), Brazil (Petrobas) and

Indonesia (Petronas) such state-owned oil companies have made forays in the international oil market.

What are then the implications of governments being identified as parties to the conflicts? First is that governments cannot and are not in a position to play the role of credible third parties in resolving these conflicts. Second, it also politicises the conflict. For instance, in the Niger Delta, many scholars and commentators situate the Niger Delta issue within the larger framework of the evolution of the Nigerian nation. Another implication is that it then ignores the relationship between the oil companies and the communities. This then gives the impression that the conflicts between oil corporations and their host communities could be legislated out of existence.

The case of Nigeria is made worse because conflict in the Niger Delta was earlier framed as resistance to military rule. And since military regimes generally lack legitimacy and are preoccupied with self-preservation they are unable to address the conflicts. Ross (2000) has tried to create a linkage between dictatorship and natural resource especially oil. He concludes that dictatorships tend to thrive under extractive industry economy like oil since they have access to quick revenue and are accountable to no one except the ruling oligarchy.

Unfortunately however the governments in Ireland have responded in exactly the same way to Shell's presence in their areas. The question then is why the similar response? The first reason is that the governments desperately need the revenue from oil to be able to carry out governance activities. Second, the oil company's as part of the globalization cartel are very influential, even more influential than governments. Third, these oil companies have their home origins in the capitals of the most influential nations in the international arena. Any sanctions or activities against these oil companies is seen as a diplomatic affront and puts a strain on the relationship between governments. Finally, the importance of oil to our civilization makes it look as if Oil Company's are sacrosanct in whatever they do.

There are other parties in the conflicts between Oil Company's and communities as in Erris and the Niger Delta. But I shall limit this analysis to this three alone. This is because some of these other parties play a role that almost always falls within the purview of these other parties. These other parties include local and international non-governmental organizations, nation states, local coalitions, the media and such international bodies as the UN, African Union, European Union and others. The important thing to note however is that all of these parties play very important part, either negatively or positively in these conflicts and can also assist in resolving the conflicts. But to what extent a government can resolve a conflict within its own borders goes a long way to showcase its legitimacy and credibility. Considering the actions and inactions of the various governments in Ireland and Nigeria, can they do much in resolving the conflicts? If they want to, are they able within the parameters of international relations do something without offending the more influential members of the international community?

Issues

One of the reasons why some conflicts are difficult to manage, resolve or transform is that interveners have not identified the most important or strategic issue or cause of the conflict. Parties in conflict use several strategies to hide or obscure their interests (Moore 1996, pp. 231-243). There are many reasons why parties engage in this. But that would be clear as the discussion gathers momentum. "Social structural and psychocultural theories attribute the primary source of conflict to different forces, each leads to distinct methods for managing conflicts successfully" (Ross 1993, p. 97).

What is important however is that for every conflict, there are three basic issues involved. These are positions, interests and needs. These three are basically different. And in analyzing any conflict and for effective resolution, this

48

clarification must be made. Every party has a position, an interest and a need. In this case the oil companies, the communities and governments all have their positions, their interests and needs. These positions, interests and needs change and evolve over time. They are not static. And as Marc Howard Ross (1993, p.22) points out

> "the structure of society is …….linked through common interests shared by individuals and groups occupying similar positions in the social structure, and pursued through collective action. Socially rooted interests can be defined in several ways. Some interests are defined through the inequalities of power and resources inherent in particular levels of social complexity. Other interests are formed through interaction and exchange among people in a society or between groups in different societies and can be described in terms of the strength of cross-cutting versus reinforcing ties."

Positions are those claims which the parties to the conflict cling on to. In Erris the position is that Shell must leave or that the gas plant and pipeline must not run through their community. In the Niger Delta one also notices the same or similar position. But the interesting thing about positions is that they change over time. For instance, in the Niger Delta, the issue was for them to enjoy the benefits of oil in terms of infrastructural development, employment and contracts. But all these have changed with the new position being resource control. And even as I write these words, the position of the people of the Niger Delta has also changed. They now want power shift. That is that they must produce the next president of Nigeria.

Even though the conflict in Erris is relatively recent, that is not as old as that of the Niger Delta; one could still observe how the positions of the parties have changed over a short period of time. Initially the people of Erris wanted to be recognized, consulted and engaged in the whole process. Now the position is that Shell must leave. They are no longer prepared to even negotiate or engage Shell.

The same could be said of the interest of the parties. In this case interests deals with those things that parties to a conflict feel that have been denied them. The unique thing about interests is that the parties could do without the interests being satisfied. Usually interests are comparative or put in other words relative. One of the interests of the people of Erris that was voiced out was that the government of Ireland was not participating in the venture. This came up when they compared the participation and owners of the other members of the consortium. In the Niger Delta as well we see that when the people compare their "participation" in the oil industry vis-à-vis other ethnic groups in Nigeria they complain. The fact about interests just like position is that they are not clear. For instance, I recall a certain case when Chevron has signed a Memorandum of Understanding with the people of Ugborodo in the Niger Delta. One of the provisions of the memorandum was the provision of a number of employment opportunities over a period of time. This was not done for sometime. But as the community got restive, Chevron appeared one morning with about 15 letters of appointment and asked the communities to supply them with names of those whose names should be written on the letters of appointment. When the community found out that the employment was to be with a Chevron contractor instead of Chevron, they turned down the offer. They have survived without the jobs and are prepared to wait until they get direct employment into Chevron. Most times, interests are more symbolic than practical.

Needs are those things which a party to a conflict must have. These are neither negotiable nor vague. Needs also do not change over time. They more or less remain constant. They are neither relative nor comparative. The life and day to day survival of the parties are involved. Without the needs being met, the people will become extinct. The people of Erris do not want any threat to their identity which is tied to their land. They want clean environment and they want safety. These needs are real. And as if to reinforce them, on September 19, 2003

> "a massive landslide drove tonnes of peat and mud off Barnacuille and Dooncarton mountains overlooking the route of the pipeline, sweeping away homes and the remains of the deceased in a nearby graveyard. The landslide covered one of the original proposed pipeline routes which, if implemented, would have been carrying unprocessed gas were it not for the planning delays" (p.16).

This very landslide has wiped away any iota of doubt that the threat of both the gas plant and pipeline are real.

More importantly there seem not to have been any serious mitigation measures put in place to counter any eventualities. The argument from Shell has been that the project is safe. In the case of the Niger Delta, they have witnessed oil spill and pollution and the poor response. There are 1,481 oil wells in the Nigerian delta. There are also 7000 kilometres of oil pipelines and flow lines. There are 275 flow stations operated by 13 different oil companies. Between 1976 and 2001, the Niger Delta has experienced 6,817 oil spill incidents. Out of this figure 6% spilled on land, 25% in the swamps and 69% offshore. About 3 million barrels of crude oil was wasted in these spills and 70% of it was never recovered. By implications the unrecovered quantity is still circulating on land, in the rivers and swamps of the Niger Delta.

They have seen environmental degradation and the non-challant attitude of subsequent governments. They have seen their sources of livelihoods poisoned, their farms rendered barren and their rivers devoid of fishes. They have also witnessed the politics of oil pollution clean-ups and how no remediation was carried out (DonPedro, 2005, p.51). Moreover, they have seen how such communities as Oloibiri was abandoned after it has been exhausted of oil and how the community is lying practically fallow from excessive exploitation. These issues are real and form the needs of the communities.

The critical issues about needs are that they are not constructed. The needs of people are real and basic. The impacts of the denial of those needs are palpable and could be felt from anywhere. However, needs may change as society evolves. For instance, a community without a school could have one built, over time the need may change from that of building a school to either adequately equipping or expanding it.

In comparing the issues involved in both Erris and the Niger Delta, the interesting thing is that both communities are at different levels of the developmental ladder. For instance, as I observed in the introduction such basics as electricity, potable water, roads, education and healthcare to a large extent has been taken care of. The people of Erris are no longer concerned with the basics; they are concerned with such "luxury necessities" as clean air, environmental integrity, preservation of identity, respect and recognition and others. It is also important that good governance is taken for granted in Ireland. Democracy has taken root over time. The citizens are enlightened and participate in the electoral process with the firm belief that their votes count.

This is not the case with the people of the Niger Delta. The people of the Niger Delta have been marginalized for a prolonged period of time. Many communities do not have such basics as potable water, educational or healthcare facilities. This is a bewildered people who have never had the fortune to experience good governance even at a minimal level. They have been marginalized, cheated and ignored over time. More critical is that the people have seen their hospitality to the oil companies betrayed in the form of non-implementation of various memoranda of understanding (MOU). They have also seen their governments disregard them and take sides with oil companies. Even the little that the oil companies have provided have been frittered away by unscrupulous leaders and government officials. In all these it is obvious that the people are going to confront these challenges from different perspectives.

However before going into responses, I will like to carefully review the various contexts under which these two conflicts are evolving. This is because Coleman & Deutsch (2000, p. 432) have argued that one of the issues that make a conflict intractable is the context. By context here is meant the milieu in which the conflict is taking place. It includes the location, time and nature of the global context. This I shall examine in order to see how the context affected the responses of both situations. This will be a short and focused discussion since I have alluded to the context since the beginning of the work.

Context

In conflict analysis, the context of the conflict is important in understanding the dynamics and structure of the conflict. The context of conflict includes the location (geography), the time (date), the culture of the area (language, religion and social organization) and the actors among other variables. The importance of the context of a conflict lies in the fact that by understanding the context, it gives us insights into who the actors are, the kind of resources being mobilized to prosecute the conflict, whether the conflict is protracted, intransigent or deep-rooted. The context of the conflict also enables us to plan for effective intervention. And to a large extent, the context of the conflict plays a role in whether the conflict will be violent or non-violent. The context of conflict also determines whether the conflict is escalating or de-escalating. As the context of a conflict changes, so does the conflict itself. For instance, as conflict plays itself out over time the actors change, so also the rhetoric and the issues. So it is within this purview that it is important that we examine and compare the contexts of the Niger Delta and Erris. To capture both the similarities and differences I shall use a table format. This will give a visual representation of what we are trying to illustrate. The themes shall be limited to ten only for brevity and to allow for space to analyze some of the issues. Of course there are other categories of comparison but we shall limit this because of space. It is also important to note that some of these categories do also change from time to time.

Table Comparing the Context of Conflict in Ireland and Niger Delta Differences

Erris	Niger Delta
North (hemisphere)	South (hemisphere)
European	African
Mono-ethnic	Multi-ethnic
Durable (stable) democracy	Emerging democracy
Comparatively affluent	Relatively poor
Developed	Underdeveloped
Christian (atheist)	Multi-religious
Powerful Diaspora population	Powerless Diaspora population
Small population	Large population
Duration of conflict (relatively recent)	Duration of conflict (more than 50 years)

The world is divided into southern and northern hemisphere. While the north is privileged the south is the under-dog. The reason for this disparity and unequal status in the world should not detain us since social scientists are yet to agree on what the causes are. The disparity and differences between Europe and Africa are too numerous to warrant recounting here. Suffice to mention that conflicts in both settings have almost always played out differently. I shall discuss this in detail when I shall deal with the responses to the conflict. But one of the most persuasive arguments for the relative stability of Europe and North America is the seeming acceptance of the unifying mission of the nation state. In Europe and North America, the nation state has been accepted as the most important unit of social, political and economic organisation. In Africa the concept of modern state is still evolving and being debated. This is against the backdrop of indigenous people's rights.

That Ireland especially Erris is mainly one ethnic group with some sprinkling of other European nationalities will definitely give the conflict a different tone. This is unlike in the Niger Delta where there are more than twenty different ethnic groups laying claim to the same resource. It is even worse in the case of the Niger Delta because the Nigerian nation itself has more than 250 ethnic groups. This means that apart from the conflict between the oil company and the communities, there are also conflicts between and within the communities.

Closely related to this is the issue of religion. In the Niger Delta, there is this mindset that it is the Muslims from the north who have conspired to deny them access to their God-given resources. This has been reinforced by the fact that out of 48 years of independence Muslims have ruled Nigerian for more than 23 years. The Niger Delta situation is complicated because apart from different Christian sects, there are also different religious sects (e.g. Muslims). But Ireland is mainly Christian namely Catholics and Protestants even though there is misunderstanding between the Protestants and Catholics. And fortunately for Erris it is not one of the places where there is tension between Catholics and Protestants. Increasingly, the influence of religion in human affairs in Europe and North America is waning. This is not the case in Africa. For instance, it has been suggested that militants of the Niger Delta rely on "juju" to make their bodies impregnable to bullets. At times one notices references to their gods in their rhetoric. All these go to reinforce the place and role of religion in these issues.

Democracies by their very nature tend to manage conflicts better, that is in non-violent manner. Moreover, democracies have developed several institutional mechanisms for managing conflicts. This is unlike in dictatorships or emerging democracies. In fact many commentators believe that Nigeria does not have a democracy yet but "civil rule." This is because some of the cardinal features of democracy are yet to take root. In Nigeria for instance, the conflicts in the Niger Delta have at various times been framed as democracy discourse. Ken Saro Wiwa was arrested, tried and killed under a military dictatorship. This may not have been this drastic under a democracy. While the conflicts among the Ogonis (Ken's ethnic group) was rooted in whether or not the Ogonis should boycott the 1993 polls. I shall also discuss this more when we discuss the responses.

There is a relationship between underdevelopment, poverty and conflict. Poverty causes conflict while violent conflict impoverishes. Though it has been argued that resource-rich countries tend to be conflicted, but Ireland's wealth could not be said to have come from mainly natural resources. Ireland has a sound technological base and a very vibrant and robust Diaspora population. This is unlike Nigeria where the presence of oil has led to a rentier economy that is not accountable. Poverty and underdevelopment has affected how the people of Ireland and Niger Delta have perceived and responded to the conflicts in their area.

The large area and population of the area of the Niger Delta have also affected the conflict. This is unlike Erris where the population is relatively small and the landmass not too wide. This is also related to the number of oil companies operating in the area. This has many implications for the communities always compare how one oil company is behaving as against the other. However, the relatively large population of the Niger Delta has led to a multiplicity of actors and issues which are almost unmanageable.

Similarities

Erris	Niger Delta
Same oil company	Same oil company
Small minority area	Small minority area
Agrarian	Agrarian
Fragile ecosystem	Fragile ecosystem
Marginalized, neglected and poor	Marginalized, neglected and poor
Strong and shared communal identity	Strong and shared communal identity
Identity connected to land	Identity connected to land
Previous colonial experience	Previous colonial experience
Not participating in oil industry	Not participating in oil industry

The similarities identified pretty well speak for themselves. The fact that it is the same oil company with a history of conflict and a reputation for environmental, social and corporate irresponsibility makes it obvious that they will not be welcome no matter their good intentions.

Both Erris and the Niger Delta are small minority and agrarian areas. Any major tampering with their environment will create massive dislocation and untold social upheaval. This is also related to their fragile ecosystem. Both areas share an alluring scenic beauty which might become a relic of mournful fascination if the oil industry which according to the Public Interest Research Group of the US, "is one of the dirtiest and most destructive industries on the planet. Onshore or offshore.......the environmental track record of the oil industry is a dirty one" (Manuel, 2001, p.3). Moreover the communities do not have any reasonable and believable assurance that if anything happens that their safety will be guaranteed.

Both Erris and the Niger Delta are marginalized, neglected and poor areas of the country. The marginalization, neglect and poverty is rooted in history. This means that issues affecting them will be seen by the people as an extension of the same oppression which they have endured over the years. And since they have a history of resisting their oppressors, they will have no qualms that they will successfully resist this one.

When a group of people as a result of shared history, now develop a strong sense of communal identity that is tied to land gets involved in a conflict, that conflict is bound to be protracted because conflicts involving land and identity have to do with value and resource. As USAID Office of Conflict Management and Mitigation asserts, "...conflict over land often combines strong economic and emotional values...it should also be kept in mind that competition over access to land is often, at its core, about power, both socio-economic and political" (p.3).

The Niger Delta and Erris share a common affinity to land and this could provide a possible explanation for their response to the conflict in their area. This goes also for their common colonial experience and the fact that both communities are not participating in the oil industry. This sense of relative deprivation according Ted Robert Gurr (1971) could explain why communities rebel.

Responses

Basically there are two responses to conflicts – violent and nonviolent. Within these two broad categories of responses are others. But I shall not dwell on the theoretical conception of responses to conflicts. I shall critically analyze and compare how the people of Erris and the Niger Delta responded to their conflicts. To my mind, the whole idea of responding to conflict is in a sense how people perceive and respond to change. To respond to conflict is to agitate for change. Third, in both instances we see people in different circumstances of life responding to conflicts of the same kind. Scholars and practitioners have described two basic responses to conflicts

(Sharp, 2005). First is violent and the second is non-violent. This categorization into two, gives the either/or mindset to responding to conflict. But a more discerning look at the issue will show that it is not either/or. Usually in responding to conflicts people start from a non-violent approach to violent. Violent approach is used when a party to the conflict usually a less powerful party is frustrated with the ineffectiveness of such non-violent approaches as negotiation, mediation etc. As one move from violent to non-violent responses, the parties increasingly lose their power to resolve and this power is surrendered to third parties.

On the other hand, a more powerful party may use violent means to get a quick result, maintain the status quo, and intimidate the less powerful party or to instigate a non-violent approach such as negotiation. At other times the violent approach may be used to save face or as James Gilligan, the respected American psychiatrist put it to gain respect. All these are dependent on several factors such as the context of the conflict, issues involved, the parties etc.

Responses to conflicts are important because it gives us insight into how parties perceive their environment, 'image the future,' and construct their 'seeing and being' in the world. But the most significant thing about understanding responses to conflict is that it provides a window for strategic and sustainable peacebuilding and long term conflict transformation. It also helps in designing structures on which just relationships are built and sustained.

Responses to conflicts are also contextual. And as Ross (1993, p.96) puts it, "effective conflict management strategies must be consistent with existing cultural norms and practices and cannot import methods that are successful in other settings without paying attention to their application in local contexts." Continuing Ross notes that "often the choice of which one to use is tied to the disputants' position in the social system" (p.96).

From the above, it is obvious that responses to conflicts are not arbitrary choices which individuals make without too much thought. Before people respond to conflicts they consider a lot issues. However, people copy from other areas especially if those responses were successful. Even where some of the responses were not successful, people still copy others if those responses have attracted attention to their causes or issues. Finally people do not respond to conflict in a specific manner because they think that it will address their issues. They respond because at times they feel that the particular response will draw attention to their plight.

I shall discuss the responses under different sub-heads or themes as the case may be.

Compensation policy

One of the major points of disagreement between communities and corporations is in the payment of compensation. There are many reasons why companies pay compensation. They may pay compensation to acquire land, to mitigate the impact of their presence or for accidents arising from their operations. In Erris the way and manner in which compensation was administered was shoddy. Moreover, it is only when a community has accepted the presence of a company that it could accept compensation. Compensations are not just palliatives, they are also tacit acceptance and consent of the community for the company to come in and do business.

In Erris let us see how Shell administered compensations. According to the Centre for Public Inquiry "people first heard of the development in spring of 2000, when local fishermen said they had been paid £2000 a piece by EEI to stay away from the area of the rig and the inshore development. During the summer of 2000, EEI began to seek support for their plans among the Erris community and donated £8,000 to the Carne Golf Club. If one interpreted the gift to the fishermen as compensation for their fishing on what basis was it calculated? Who negotiated it? And who owns the sea on which the fishermen fished?

Another instance of compensation in Erris is when the consortium approached the residents to implement the CAOs. According to the CPI, "within weeks, landowners along the route of the proposed pipeline were informed that they would be served with CAOs unless they accepted compensation and allowed EEI to lay the pipeline" (p.14). Here compensation was accompanied with threats and blackmail.

As I alluded to before, compensation is loaded with meanings. As Micheal Seighin put it, "the farms form the basis of the identity of the people, monetary compensation cannot compensate for undermining the social identity of the people" (Connolly & Lynch, 2005, p.45). As illustrated by the people of Erris, compensation often do not consider the interests of the next generation or the future losses which a community might incur. Compensations must be managed in such a way that the process must be transparent, respectful of the people and sustainable in conception. The amount of compensation must also be such that is acceptable to most if not all.

The situation in the Niger Delta is similar in a way and different in other aspects. The governments of Nigeria do not require CAOs to acquire any piece of land. The Land Use Act of 1978 has already appropriated land for government. Second in the Niger Delta compensation is paid for buildings and crops and not for land. The issue in Nigeria has been the inadequacy of compensation. The laws has remained unchanged for close to 50 years.

This is also made more complex by the complicated land tenure system prevalent in the Niger Delta. There is individual, family, community and government ownership of land. The question is then who gets the compensation? At times when compensation is paid, governments even want to control it for either the community or the individual. A good case in point was when Chevron paid compensation to Ugborodo community and the Delta State Government wanted to use the money as counterpart funding for a housing project.

In some other instances Shell has been accused of "deceitful acquisition of land" (Vanguard, Friday, July 28, 2006, p.6). In this case the Delta State House of Assembly asked Shell to pay the Ikpereguma Family the sum of #200 million (approximately 1.3 million Euros). All these lead to conflicts.

Managing community expectations

Whenever a company, more so an oil company appears on the horizon, there is heightened expectation of possible benefits. This was the case in Erris. Let me allow the people of Erris speak for themselves. As CPI put it "initially, locals were excited about the prospect. We thought we would have gas coming into County Mayo, a respondent who owns land along the pipeline route was quoted as saying. Another said "Our son was working up in Dublin at the time, and he and other people thought that they would have jobs to come back to". Construction of the pipeline and the gas processing plant promised the creation of up to 500 local jobs in an area with an unemployment rate of more than 30%. At that point residents assumed the pipeline was to carry clean gas but soon learned the distinction between upstream and downstream pipeline. The upstream pipeline will carry unprocessed gas, which contains a volatile mix of chemical compounds, from the subsea field to the gas processing plant while the downstream pipeline carries clean, processed and consumer ready gas" (p.26).

Shell staff marketed these benefits to the people of Erris. Was it a deliberate ploy to deceive and gain entry into the community? Or was it a case of situated optimism of the optimal value derivable from the investments? Another question is this: are the above benefits derivable from the project? There are precedents especially from Norway and Britain countries bordering Ireland. More importantly it was not the first time that Ireland was dealing with oil companies. Since 1978 Marathon Oil has been supplying natural gas to Ireland from the Kinsdale gas field. Reports indicate that the Kinsdale gas field has created hundreds of jobs and made gas available at reasonable price.

The presence of oil companies is seen as a saving grace from the neglect of governments especially in the Niger Delta. The case of the Niger Delta is made more complicated since the people of the area are minorities in Nigeria's political configuration. But consistently, the people have not derived the expected benefits from oil exploitation. This has been one of the grievances of the people of the Niger Delta. And they have responded by insisting that the oil companies should leave or that they be given a better deal.

The interesting thing to note here is that the oil company amplified the derivable benefits as their response while the disappointment of the community led to protests and demonstration.

Litigation

Increasingly scholars have queried the use of litigation as a conflict resolution mechanism (Zehr, 1990; Rothman, 1997; Kriesberg, 2003, Frynas 2000, Onuoha, 2007, etc.). This is especially where long term relationship is envisaged. Litigation has in most instances failed to address relational issues. But in our case study

> "when five men refused Shell access to their lands in January 2005, Shell obtained a High Court injunction against the men and the five were locked up for more than three months drawing national and international attention to their battle against the oil industry" (p.22).

Instead of imprisonment resolving the conflict, here is how the residents of Rossport responded to the incarceration of their fellow citizens,

63

"we remind Shell and their Irish government partner that imprisonments have historically and will always fail as a method to secure the agreement of Irish people. We now call on our supporters to intensify the campaign for the safety of our community and families, the campaign has now begun in earnest" (p.49).

Two observations are important here. First, is that the Rossport Five after the jail did no longer see themselves as mere resident of Erris. They reminded whoever cared to listen that imprisonment "historically and will always fail". They linked the past with the future and this has great implications for the existence of the company in the area. Second, the imprisonment was used to remind the people of Ireland that Shell is not indigenous and that Shell's relationship with the British even though Dutch is a reminder of more than 300 years of colonialism in Ireland. The Northern Ireland issue also reverberates. As one respondent told me, the government is being careful because whatever happens here may have great implications for the peace process in Northern Ireland which is still under British rule. By imprisoning the Rossport Five Shell unwittingly made martyrs out of regular citizens. And as they declared; "the campaign has now begun in earnest" (p. 49). In other words litigation instead of ending the conflict heralded its beginning.

In the Niger Delta the oil companies and communities have at one time or the other used litigation as a response to conflict. Ordinarily this should be commended. But it is not as simple as it appears. Frynas (2000) has shown why communities in the Niger Delta may never get justice from litigation. Second, he says that communities in the Niger Delta do not possess the wherewithal to take on oil giants like Shell. This reminds us of Ken Saro Wiwa who was killed ostensibly through a judicial process. There have been other instances of the use of litigation. For instance, the people of Umuechem went to court against Shell, the case was stalled. They even went to the National Assembly committee on public petition still nothing happened. They went to the Human Rights Violations and Investigations Commission (Nigeria's equivalent of a Truth Commission) and nothing happened. As I said earlier, it is the frustration from these responses that lead to violence. If litigation has failed to address community issues especially as it relates to their relationship with multinational corporations, why then have the parties insisted on using it? There are so many reasons for this, but the most important is that corporations like to see themselves as being law abiding.

Impunity

Closely related to the issue of the use of litigation is that of impunity. By impunity here I mean how companies operate as if they are above the laws of the land. I also mean a selective use and application of the laws of the land by especially oil companies. On several occasions in this case study the company was caught violating the law. For instance, Shell was accused of building a pipeline without ministerial consent. To substantiate this, the Pro Erris Gas Group asked that Shell pay the community $250, 000 instead of dismantling the pipelines. In its report CPI noted that

> "in July, it emerged that Shell had breached its consents by welding together three kilometres of pipeline at the Ballinaboy site. It also emerged after residents raised the issue, that the PAD, which has been charged with monitoring the project, had relied on regular reports from Shell rather than direct site visits" (p.47).

I was also shown pipes already welded by Shell at their site without authorisation. Another case in point was in October 2000 when Bertie Ahern (a minister) announced that the "consortium would build a connector pipeline from the Ballinaboy site to the national loop at Galway. This announcement was made before any application for planning permission for the project was submitted by the developers" (p.29).

It is pedestrian to say that he who must go to equity must go with clean hands. What the people of Erris found unsettling is the fact that an organisation that found it convenient to use the instrumentalities of the law to pursue its own ends found it difficult to obey the same laws. And moreover the company wants the residents of Erris to comply with Compulsory Acquisition Orders of dubious and doubtful origin.

This issue of impunity has also been responsible for various accusations of human rights violations against oil companies. In Erris this is at the heart of the conflict.

The case of the Niger Delta is made complex by two related factors. First, is military rule which has held Nigeria down for a long time. Oil companies have also perfected the art of blackmailing governments. Second, is the excessive reliant of governments on oil revenue. Because of these oil companies operate with impunity and the communities have no response to this except to try self-help. Even where oil companies want to be decent they engage in delay tactics and unnecessary appeals to hide their impunity. So in Erris and Niger Delta oil companies have shown a total disregard for the rule of law.

Stakeholder engagement process

Stakeholder engagement process could be interpreted from two perspectives. First, is how the company consults with the people before making decisions, especially those that affect the community. Second is how the company manages relationships with the community. As the Collaborative for Development Action (CDA) puts it

"there are two types of processes that companies most frequently use in working with stakeholders: consultation and negotiation. Broadly defined, a negotiation is a process of meetings deliberately convened to reach agreement on a particular issue. A consultation process is a more open-ended

66

set of conversations or meetings, with the objective of exchanging ideas and opinions (without formally coming to an agreement)" (CEP, p.2).

On their web site Shell said that it values local opinion, support and input as critical to the success of the project. For emphasis, Shell Ireland spokesperson Susan Shannon said, "we have always tried to communicate with the local stakeholders." But one of my respondents disagrees. Shell he argued has been working with a group known as the Pro Erris Gas Group which is supporting the project. In fact it was this group that asked Shell to pay the community the sum of €250,000 rather than dismantle the pipeline which was built without ministerial consent.

According to one of my informants Vincent McGrath, one of the Rossport Five,

"my first grouse is that in all these Shell never for once contacted us who live on this pipeline way. The first time I heard from Shell was when they sent a letter informing me that the pipeline will run 70 meters away from my house. And that was a safe distance. How they arrived at that figure I do not know and may never know. The second time was when they committed me to court and sent me to jail. Is that how a responsible organization should behave in the 21st century?"

During my visit to Erris I watched a video recording of the encounter between Shell staff and helpers and the community. It was at the Community centre where Shell has invited members to discuss issues arising from the pipeline. Some people were not allowed to participate in the meeting. Though they eventually found their way into the meeting venue, community members felt violated by the way they were treated at the meeting.

Another important aspect is how the company selects those to consult with. In the case study there are two groups. Shell decided to support one group against the other. Even though it is natural for individuals and organisations to bond with those that tend to support them, it is also important to be careful in selecting those to be consulted. As one of my informants put it, "On Wednesday Shell sent out an invitation for a meeting. They said that it was a "general invitation". When we arrived at the meeting, they were now screening those who will be allowed in and those who will not. Who are they deceiving? Stakeholders for Shell are those that help their business." I also had a similar experience when Shell was putting in place mechanisms for the Global Memorandum of Understanding (GMOU) in Nigeria. When we arrived at the meeting, we were practically worked out. I apologised and left. I also requested to meet with the manager in question. A week later, I wrote her an email requesting a meeting and to monitor the GMOU process. As I write I have never received any response from her. When I saw her at an interactive session with people from United States Institute of Peace (USIP) and *Search for Common Ground* (SFCG) she pretended not to have recognised me.

The way and manner that Shell went about asking people to comply with the CAOs and promising compensation were all part of their community engagement process. Obviously, that method did little to build their relationship with the community. There are still other instances where community members felt that a more pro-active and integrative community engagement process would have paid off.

Local content

From the interviews and my interactions with the people of Erris I could not establish what the requirement was for local content. For instance in Nigeria I am aware that it is about 70%. By local content here I mean to what extent the local population is involved in such company activities as contracting and employment. This poses two dilemmas. First, is that of number (in the case of employment) and second is that of amount and type (in the case of contracts). In one word local content to me simply mean benefits which the community could derive from hosting the company.

From the case study this raises a lot of issues. First, since the Republic of Ireland is not participating and the oil company has been granted generous tax concessions, it is obvious that the people interpreted the local content as being virtually non-existent.

Second, the cheap gas which they expected from the project was not to be. This posed another dilemma of local content. Contractors welding the pipelines in Erris are from Italy. I could not ascertain the other companies working for Shell in the area. More importantly, instead of improving their local content the project will take away 400 acres of their land and a long mileage of pipeline with very high risk of accident and pollution.

The people did not see any benefits that would accrue to them for hosting the project and they made it clear. This was in spite of the fact that both the government and the oil company have sold the project as being immensely beneficial. Another aspect of local content that is worth considering is even if there have been employment of the locals, at what levels were they hired? If they were awarded contracts, what is the value of these contracts? And to what extent are they able to access those managing the company to be able to discuss issues especially when they notice violations of local content regulations. And finally who enforces or takes the responsibility for enforcing violations of local content regulation?

In the Niger Delta the issue of local content is a little more complicated. Oil companies enter into joint venture agreements with the federal government. This means that local content policy must be interpreted within the framework of the Nigerian nation. Let me illustrate. If for instance, the local content policy stipulates 70% employment that means that 70% of the jobs must go to Nigerians. But this is not how this is interpreted. The 70% is interpreted in favour of the specific communities where oil company operations are domiciled. At the core of this interpretation of local content is ethnicity and history. For instance, when other regions of Nigeria had agricultural produce they were used exclusively for the benefit of the regions. This underscores the point which was made earlier about the context, location and time of conflict as being important in understanding the dynamics of the conflict.

Relationship with government

The relationship between oil companies and governments has been a major area of concern to researchers of government-corporation relationship. In the oil industry this has been more so because of the peculiar nature of the oil industry. Oil is a strategic resource which wherever it is exploited, governments have played a major role. The Voluntary Principles on Security and Human Rights as enunciated by the participating members have clearly specified the modalities for relationship between the corporations, communities and governments. The Voluntary Principles envisages a tripartite relationship anchored on mutual dialogue.

In the case of Erris there is this widespread believe that the relationship between the Republic of Ireland and Shell has been too cosy. In other words, it is too lopsided in favour of the corporations thereby leaving out the communities. My informants point to the speed and the number of amendments to several legislations just to facilitate an easy entry for Shell. In fact CPI (p.15) quotes the Department of the Environment as promising that "all possible steps will be taken by the board to ensure that any such appeal is processed with all possible speed with a view to giving a final decision on it within the statutory objective period of 18 weeks." This has led to allegations of corruption. In this case study one government official was found to have received payments from the oil company.

However, government officials have often explained away the generous terms as part of the policy of attracting the so-called foreign direct investment. For instance in 1987, Ray Burke who took over the energy ministry announced the exemption of oil and gas production from royalty payments (p.11). He explained this away as being "necessary in the light of the poor drilling results of previous years and the low price of crude oil" (p.11). In 1992 also the government introduced new licensing terms that allowed oil companies to sell at market prices instead of the earlier arrangement of "bulk discount agreement."

But what peeved the people of Erris the more was Statutory Instrument, SI 517 of November 2001 which gave the minister power to make Compulsory Acquisition Orders and in March 2002, this was incorporated into the Gas Act. This is significant because this particular legislation allowed private companies to enter private land and build oil facilities whether the landowners agreed or not. According to one of my informants

"You see, the meaning of this is that a private company now has the right to take land from people whether they like it or not. However, they were to be paid some kind of compensation. If they rejected the compensation, the land will be taken anyway. This has never happened in the history of Ireland before. Before we understood what was going on we were served with the Compulsory Acquisition Orders (CAO) unless we accepted compensation or allowed the company to lay the pipeline. Still there were further amendments to these orders".

Another area that affects relationship is the use of security forces. Police or the military are usually deployed in conflict situations. The police especially represent the state's own instrument of coercion. A group of people who have been marginalized in a given polity will see the deployment of troops in their area as part of continuing oppression. It also signifies that the company's intentions are unwholesome. As one of my informants put it, "How can a business go about with the gardai (police)? On the other hand companies have responded that their use of security forces are either routine or to protect their properties. However, what is missing is the meaning which people attach to the use of military personnel. This makes the communities to equate the companies with an army of occupation.

One of the most enduring complaints of the people of the Niger Delta is the seeming inability or unwillingness of government to rein in oil companies. At every point in time, the government has always been too eager to patronise and protect oil companies. On the other hand also, the people of the Niger Delta have decried the militarization of the area. For instance, there is the Niger Delta Task Force on Security. This task force is made up of the Army, Navy, Air force and the Police. Space will not allow me to go into details as to the origin of the task. But suffice to mention that after one of the clashes between two ethnic groups in the Niger Delta, they called on government to establish a task force on security to monitor the area. After the task force was set up, it no longer serves the communities but is financed by oil companies. This task force has been accused and implicated in several human rights abuses (see the documentary "Fence Too High" produced by the Centre for Social and Corporate Responsibility based in Nigeria).

Integrity of consultants and contractors

Because the oil industry is a technical one it makes extensive use of specialist consultants. And because oil operations can be quite hazardous these consultants must be credible. The same is applicable to the use of contractors. But when community members begin to receive mixed signals as to the integrity and professional competence of consultants and contractor, it raises issues of community relations.

In Erris many aspects of the project have attracted the use of consultants and contractors. First was in the development of Quantified Risk Assessment (QRA). According to CPI (p.33) "QRA is a mathematical modelling tool used by engineers to quantify risk to human safety." This document has been mired in controversy since the commencement of the Corrib Gas Project. J. P. Kenny produced the first QRA in November 2001. This has not been released ever since. Apart from this five other versions have been produced and these are still to be widely accepted.

Another angle to the use of expert opinion and advice is the relationship with oil companies. In one instance it was found that Shell owned 60% equity in one of the consulting firms that have been used. In another instance, the consortium has commissioned a report from the Coastal and Marine Resources Centre at University College Cork. The report stated that, "Broadhaven Bay SAC and its neighbouring coastal waters undoubtedly represent an important area for marine mammals and other species." However in Shell's Environmental Impact Statement which was submitted to Mayo County Council, it stated that there is "no evidence that the bay is of particular importance to whales and dolphins" (p.27).

More worrying to the people of Erris is also how Shell arrived at the 70 metres safe distance of the pipeline away from dwellings and shelter. It is worrisome because this is the exact distance from the home of Vincent McGrath. Moreover, as McGrath put it,

"The first time I heard from Shell was when they sent a letter informing me that the pipeline will run 70 meters away from my house. And that was a safe distance. How they arrived at that figure I do not know and may never know. ....You probably heard of the Carlsbad pipeline explosion in the US. People living 400 metres away were burnt to death and they are telling me that 70 meters away is safe distance."

Reconciling this kind of imposed finding is an issue. And some of my informants complained that they were not even allowed to see the reports. Others insisted that no independent assessments have recommended the project. All these call to question the integrity and competence of Shell consultants and contractors. It also raises questions of the independence of assessments by consultants and contractors retained by oil companies.

The situation in the Niger Delta is slightly different and worrisome. Before the commencement of any project, oil companies are expected to conduct environmental impact assessments (EIA). More often than not these EIAs are hardly made public that is if they are conducted. People of the Niger Delta hardly have access to such assessments and reports. The main reason for this is that oil companies feel that the communities do not have the capacity and capability to understand the content of such technical reports.

A more disturbing phenomenon is the falsehood propagated in some of these assessment reports. This is mainly prevalent in cases of oil spill and other environmental issues. The impression is given that proper clean up of the spill has been done and remediation carried out. A visit to the sites shows the contrary. It is interesting that in Ireland people have access to these reports and even queried them, in the Niger Delta it is the exclusive preserve of some anointed elites and oil company cohorts. Increasingly civil society organisations have taken it upon themselves to build the capacity of communities to be part of the EIA process. In the Niger Delta the Environmental Rights Action (ERA) has done so much work in this regard.

Choosing a location

What criteria do oil companies use to select locations for their projects? Is it cost (euphemism for profit) or safety and sustainability? In the case of Erris an alternative location was suggested for the project but this was turned down by the developers. In September 2003 there was a massive landslide which tore through part of the area where the pipeline was to run through. Apart from calling to question the competence of the company's consultants and contractors it also raised doubts about the intentions of the company it choosing its project sites.

Moreover, a major plank of the Shell-Sea-Campaign was that Shell process the gas on a shallow platform. The company refused claiming that it was unviable. As Moore a retired planning inspector noted

> "from a strategic planning perspective this is the wrong site. From the perspective of government policy which seeks to foster regional development, this is the wrong site, from the perspective of minimising environmental impact, this is the wrong site, and consequently, from the perspective of sustainable development, this is the wrong site" (p.35).

If there is this overwhelming evidence of the unsuitability of the site, why insist on it? Analysts have pointed to the need for profit maximisation, while others have argued that it is human pride and ego.

In Nigeria, choosing a site for business has lots of implications. First, because Nigeria is a federal state, business location means more revenue for the hosting federal component. It also implies more employment and contract opportunities for the people of the area. But a more interesting finding is that businesses in Nigeria arbitrarily choose locations or obey government directive. For instance, Shell has to move its headquarters from Port Harcourt to Lagos based on a federal government 1968 directive. Chevron has its operational headquarters in Lagos while its oil fields, flow stations and export terminals are located in the heart of the Niger Delta at Escravos.

There are two issues involved in choosing a business location. First, is the safety of the communities and sustainability of the project. Second, is cost such as proximity to the source of raw materials. And third is political expediency. The point is to what extent are host communities involved in this process? Or should they really be involved? Or what should be the overriding consideration in choosing a business location, profit, safety or political expediency? Whatever be the case it is important to bear it in mind that businesses are set up to solve people's problems, if businesses then turned round to create problems for people, then that business decision needs to be revisited.

Corporate social responsibility policy

Corporate Social Responsibility (CSR) has in the last decade become an important aspect of the curriculum of Business Schools. The argument is that should CSR be mere philanthropy and public relations or should it be part of corporate policy for sustainability? CSR has a lot to it especially in an industry that maintains high profitability with hazardous consequences. By CSR here I mean how and in what way does a company put something back into the community where it is located?

In Erris Shell gave €8,000 to the local golf club, they gave €2,000 to each fisherman to keep off the pipeline route. They also flew the Bishop and his priest to the platform offshore to bless it. Do these acts of charity constitute CSR? Can it build sustainable peace and relationship between an oil company and its host community? From the response of the people of Erris the answer is no because these are acts of tokenism or put blandly – bribery.

Their grouse with these acts are many. First, negotiations were still going on when they started distributing their largesse. Second, what criteria were used to dispense these goodies? Third, who are the beneficiaries? What the opposition of the people of Erris point to is the fact that CSR is a process and not a thing that is delivered. For any act of CSR to be meaningful it must be administered respectfully and in a transparent manner. It must not be used to divide the

76

community or where a community is already divided to deepen the division.

CSR must also be long term and planned in such a way as to be a part of the project. A critical question however which challenges researchers of CSR today is whether it should come before, with (i.e. during) or after the project. That question is not within the purview of this case study. Suffice to mention that whatever the timing for administering CSR, it must be such that must be systematic, authentic and original. Finally CSR is not only a matter of being good hearted but it is also a hardcore professional activity that requires critical thinking, in depth analysis and coordinated delivery with follow-up evaluation of impact.

It is important to note however that all the issues that have been discussed in this case study constitute aspects of CSR. CSR is an engagement process – precisely how organisations go about their community relation issues. But the most important thing is that the people of Erris think and believe that Shell does not have any CSR policy for the area and even if they had one it was not strategically conceived and that it was half-heartedly and half hazard in implementation.

What one notices in the Niger Delta is that CSR has been reactive instead of proactive. Second, CSR has been used to appease those that appear more violent and aggressive. One has also noticed the evolution of CSR in the Niger Delta from philanthropy to partnership and now to sustainable community development. In the Niger Delta, CSR has been used to respond to increasing incidence of violence, vandalism and hostage-taking. But the other side of CSR has been ignored such as prompt and accurate payment of taxes, respect of agreements between communities and oil companies and the minimisation of corruption.

Oil companies have advanced many reasons why they have a mangled CSR policy. The most prominent being that they are basically a technical organisation and not a development agency. This begs question because close to 80% activity of oil companies are contracted out. I think that CSR could also be contracted out to specialist agencies. On the other hand, such sundry sub-heads as security, payment for chiefs, surveillance contracts for restive youths must not come under CSR. This is for the simple reason that they are not.

Finally, another area of CSR that is usually ignored in the discourse is how oil companies treat their staff. In Nigeria, oil company staffs are among the best paid. But oil companies have also perfected the act of using contract or temporary staffs. They do this in order not to incur the liability that would come if the staff were permanent. And it is important for oil companies especially in the Niger Delta to spend their CSR budget on CSR!

Attitude of staff

Organizations are made up of and managed by people. In some of my interactions with some oil company staffs, I have walked away with the impression that what flows in their veins is not blood but water. And in conversations with the people of Erris one may be tempted to dismiss some of their issues as trivial. That would amount to intellectual carelessness. This is because a closer scrutiny of these issues will give us insights into the meaning of how they construct their worldview. As one of my informants reported, "the other day I was driving and a Shell staff blocked the road with his jeep. I waited for 15 minutes before he moved the vehicle. There was no apology, nothing. Is that the kind of people that we want?" When the same respondent was driving me to the trailer we saw a Jeep, she said, we don't like those here, pointing to the jeep. I was curious I asked why, she said Shell people drive them.

This very simple statement is passing on several messages. First, is that there is nothing wrong with a jeep, there is everything wrong with the driver and how he/she behaves on the road. Second, Erris as I mentioned in the introduction is a rural community where people feel a peculiar sense of self-fulfilment. It is possible that jeeps represent to them the unbridled consumerism that has made the world so greedy for oil that they are ready to endanger livelihoods and the environment for oil to flow. It is also possible that to the people of Erris jeeps symbolize the unnecessary exhibitionism that is associated with oil company staff.

The issue then is what level of cultural sensitivity training does oil company staffs undergo before being posted to a new location. What level of local anthropology are they exposed to as they mobilise for new areas? Increasingly these are issues that businesses that would be sustainable must confront in a globalizing business environment. People drive businesses, people manage businesses and people patronize or prevent businesses, CSR must also focus on people's attitude.

The use of police has been discussed before but it also has a lot to do with the attitude of staff. If one is behaving well, there is no need for police escort except may be where crime rate is high. This could not be said of Erris. The only possible explanation is as we say it "if a sheep urinates on its sleeping pen, it will prefer standing." This might be the case with the attitude of Shell staff in Erris. If that is the case, then it means that there is something peevish and even perverse about it all.

While researching my other book on the Niger Delta, one of the most common complaints of the people is the lack of respect by oil company staff. In my work I have experienced this over and over again. There is something about oil company staffs that tends to make them disregard and disrespect others. In one of my fine moments I have tried to investigate this phenomenon. One of the most recurring responses which I got was that people are jealous of oil company staffs. This may be so but does it explain their disdainful attitude towards others? Do they know and appreciate the reputational cost of their attitude to their organisations.

This attitude has several strands and implications. It has made oil company not to take community issues very seriously, it has made them to give wrong impression to their management about the actual situations on the ground, it has made them not to leverage their influence to the betterment of their communities. It has made them to misappropriate CSR budgets for other purposes other than CSR.

Both in Ireland and Nigeria this has been at the root of most conflicts between oil companies and communities. It is important because more often than not these oil company staffs are members of some of these communities. And if their attitude to their host communities are not better managed, then it would be almost impossible to build the kind of healthy relationship required for sustainable and profitable business. This is an issue which most corporations have ignored. In Ireland and even on the web site of Shell there was little or no mention about the attitude of staff. In the Niger Delta, it is also ignored as a non-issue. But it is the issue. This is because those that will interact with community members will need to develop positive and respectful attitude especially when organising and attending meetings, consultations and even in their use of words in letters to the communities.

Culture, tradition and symbolisms

In the hard-boiled business world, the issues of culture and symbolisms have been largely ignored. This is obvious. But recent developments are re-directing us to culture, traditions and symbolisms. Any company that wants to be sustainable and operate in an environment of peace must accord these issues some attention.

In Erris the people's identity is tied to their connection to the land and sea. Their farms and homes represent who they are and as Seighin (one of the Rossport Five) puts it, "no amount of compensation can pay for that." The people of Erris are proud. They are proud of their land, their culture and their heritage.

They consider it a threat to their identity if their environment and livelihood is endangered. So to them, the gas processing plant and the pipeline is not the best.

In front of Shell facility in Erris, the people erected nine crosses. On each of the crosses is the name of the Ogoni Nine. My guide having been briefed that I am a Nigerian decided to explain why they hung the crosses there. The crosses are a declaration of their Christian faith. That the people agreed to use the cross is a symbol of their unity of purpose and the importance of their commitment to non-violence. He told me that they wanted to remind Shell that they know all about them and their activities all over the world. Second, that they were not prepared to let what happened in Nigeria to happen here.

According to my informant, the trailer which I described in the introduction represented the sense of community of the people of Erris which the Corrib Gas project wants to destroy. The daily rendezvous at the trailer is a message that the people are determined to stay together as a community. These are important messages for oil companies. They only ignore this at their peril.

Human beings are meaning-making. Our lives revolve around the meanings we make or attach to events in our lives. We also make meanings of events in other people's lives and how they may affect us. Oil companies tend to disregard or even denigrate meanings and symbolisms in people's lives. In the Niger Delta, the rationalisation was that oil company staffs were mainly westerners who do not share meaning with Africans. Second, some of the culture, symbolisms and traditions were dismissed as fetish. Increasingly it is becoming clear that the main motive for disregarding people's culture, traditions and symbolisms is management recklessness. This is closely linked to the attitude of oil company staff especially in the way and manner they interpreted and related to community members.

Chapter 3

Deductions

Lessons Learnt

For every activity or intervention there are lessons for people especially activists and policy makers to learn. Learning lessons from every human endeavour enriches us as human beings and also prevents a re-occurence of the same phenomenon. There are lessons for governments, communities and oil companies.

Government

The first and most important lesson government both in Nigeria and Ireland is that citizens are becoming increasingly conscious of their right to participate in decisions that affect their lives. The era of omnipresent, omniscient and omnipotent government is gone. Governments especially democracies must thoroughly engage the people in all issues that concern them. When in 1986 the UN defined development as the felt need of the people, it meant that the people have the right to determine what constitutes development to them.

Another important lesson for governments is that there is the need for greater openness and transparency in government activities. Governments can no longer afford to run as secret cults. This calls to mind what constitutes the people and government interest. More often than not political leaders have equated their interests with that of the state. This should not be so. The people know what they want and should be given the opportunity to decide how, when and from whom to get it.

Moreover, the whole idea of extractive industries being the business of multinationals and nation states requires a strategic rethink. The people who bear the brunt of extractive industry activities must also take part in negotiating the terms and conditions for extracting their wealth. Finally governance is not all about money. Money is important but planning and prudent management is more important. The desperation with which various governments give in to oil companies must be further investigated. There is too much of international linkages in the oil industry which leaves the local communities disempowered and frustrated. The industry must be indigenized and localized.

Finally governments must put themselves in situations where they can act as credible third parties to mediate conflicts between oil companies and their host communities. The government of Ireland came to this realization a little late in the day when they appointed Peter Cassells (see Corrib Natural Gas-a presentation by Shell, 2006) as mediator. It took a long while for the Rossport Five to agree to meet with him. In the Niger Delta, the Commission of Nobel Laureates is also intervening in the conflicts between the oil companies and host communities.

The main lesson here is that government has lost the moral right, legitimacy and credibility to act as credible third party intervener. This should not be so. Government position must be such that at all times both the citizens and corporation could call on it to resolve disputes in an impartial manner.

Governments must stop using the easy but hurtful means of generating revenue at the expense of the well-being of their citizens. There are other creative ways of raising money for governance. The desperation with which governments chase after oil revenue calls to question their desire to be answerable to their people.

Oil Companies

Power intoxicates, absolute power intoxicates absolutely. Oil and gas are the two energy sources on which our civilization revolves. The oil industry is the most profitable industry in the world today. Oil companies are the most influential multinational corporations in the world today. Oil company workers and executives are all too aware of all these and have used their power and influence recklessly.

The first lesson for oil companies is that it shall be no longer business as usual. Oil companies can no longer hide under the guise of "we are not a development agency" and perpetrate gross human rights violations. Because of the nature of the oil industry and its negative impact on our environment, utmost care and caution must be taken in the conception, planning and execution of projects.

From Russia to Rio De Janeiro, from Erris to the Niger Delta, oil companies have again and again proven that they cannot be trusted both in their relationship with communities and even in their business activities. This is in spite of several guidelines, codes and frameworks that have been developed to regulate the industry. The big lesson is that there is a clear need for restructuring the training which oil company staffs receive.

Oil companies must learn to build trust and confidence and minimize mistrust and suspicion. To build trust and confidence, oil companies must undertake the following:

- ✓ Before and at the commencement of any project, oil companies must engage in early stakeholder engagement and institute stakeholder engagement processes.
- ✓ Stakeholder engagement must be consistent
- ✓ Stakeholder engagement must be meaningful and not for PR purposes
- ✓ Stakeholder engagement must be empowering to all
- ✓ Oil companies must be transparent about their plans
- ✓ Oil companies must be open and transparent about their schedules
- ✓ Oil companies must clearly communicate their prospects to the stakeholders
- ✓ Oil company staff must be respectful in all their dealings with their stakeholders

✓ Oil companies must create and maintain effective channels for raising and addressing issues (see Stakeholder Research Associates Canada, Inc. 2005).

Finally companies must build the capacity of their stakeholders and not take advantage of their ignorance. This is because whenever the stakeholders suddenly discover that they have been cheated all along, no one could guarantee their response. This has happened in the Niger Delta until Ken Saro Wiwa mobilized his people and ever since the Niger Delta has not been the same again. More importantly, this opened the eyes of other communities and they began to fight the oil companies.

Again as International Alert (March 2005, Section 4, Flashpoint Issue 1, p.4) puts it "companies should begin engaging early with communities and other stakeholders even before exploratory operations commence. Some begin the process as early as one and a half to two years prior to exploration." This has a lot of implications for building confidence, trust and partnership. Statoil in Nigeria has started early enough with the establishment of the Akassa Development Foundation in Bayelsa State of Nigeria.

"Shell apologises for jailings" is the creaming headline of Corrib Gas Update, Issue number 2 of May 2006. It need not have come to this. In Ireland and the Niger Delta, Shell and in fact oil companies generally have through acts of omission or commission injured members of the local populace. In Ireland the jailing of the Rossport Five is like a premeditated public relations disaster. In Ogoniland the killing of the Ogoni chiefs and the subsequent execution of Ken Saro Wiwa is a burden that Shell and the entire extractive industry shall carry for a long time. These acts are clearly avoidable. The personalization of corporate irresponsibility is a trauma that even when healed is almost impossible for community members to transcend and oil companies to overcome.

Communities

The first and most important lesson for communities both in Erris and the Niger Delta is that communities must be prepared and ready before the arrival of the guest with hunch back. This is made more complex by the increasing reluctance on the part of governments to be responsible and responsive to the felt needs of their people.

The lesson from Erris in Ireland is that the people have taken it for granted that all is well and that they may not have a need for a community platform to engage issues. So when Shell arrived they have to put together a community forum namely the *Shell-to-Sea-Campaign*. This body is ad hoc and lacking in any clear cut leadership model is prone to being disbanded when the Shell issue dies down. I remember asking one of my informants what will happen after Shell. She looked at me, shrugged her shoulders and said, "I don't know."

The innovative organizational model of the people of Erris must be emulated by the people of the Niger Delta. The spirit of volunteerism is very high among the people of Erris. I did not see any instances of leadership tussle since every member is a leader of sorts. Elsewhere (Onuoha, 2005) I have advocated for the intellectualization of the struggle of the people of the Niger Delta. This came out very clearly in Ireland. As reported by the Centre for Public Inquiry (p. 17) it was found out that a consulting firm, which everyone thought, was an independent and professional consulting firm was partly owned by Shell. This intellectual ferment is lacking in the Niger Delta. The oil industry values its reputation, in order to get their attention that reputation must be confronted with facts.

The most important lesson for the communities especially in the Niger Delta is the non-violent manner in which the people of Erris have conducted their campaign. I am aware that the manifestations of violence in the Niger Delta are the outcome of frustration. It will be of much value for a further investigation of the strategy of the people of Erris. When I asked many of the campaigners, they told me that non-violence was not a deliberate choice which they made but that it was almost like a natural part to tow. They also told me that some people have also tried at times "to go over the brink" such have been checked by others.

The criminalization of activism in the Niger Delta is one lesson the people of Erris must be wary of. This has been caused mainly by the infiltration of mercenaries and the proliferation of small arms. This to a very large extent has complicated the struggle of the people of the Niger Delta. It has also at times denied the people of the Delta region the much needed support and goodwill of the international community. For instance, the employment of such tactics as kidnapping and demand for ransom, bombing and the violent storming of oil company facilities clearly shows a copying of terrorist tactics from Iraq and elsewhere. And in our present world where terrorism has become a challenge to the international community, anyone or group that is labelled as a terrorist will clearly not find sympathizers for their cause.

Concluding thoughts on the Court Case

The men (Rossport Five) were released from Cloverhill Prison on September 30, 2005. This was exactly 95 days after their incarceration. It is interesting that about 30 others who were with these five men were never charged. Andy Pyle the CEO of Shell Ireland was quoted as saying that, "the fact is that we've gone through a process, and we have five people who don't like the outcome." Going by media reports, it is not only five men who did not like the outcome. In fact a survey in September 2006, approximately one year after the men were released from jail, by TNS/MRBI commissioned by RTE found that, "two thirds of those surveyed (throughout Mayo County) supported the stance taken by the five men from Rossport in their defiance of a court order in relation to the Corrib gas project."

There are some interesting observations and parallelisms of this court case and issues in the Niger Delta. It is noteworthy that Shell sort for and got the injunctions of which it was in contravention that the men were charged for contempt. I do not recall any instance in Nigeria when someone has been jailed as a direct result of litigation by an oil company or specifically by Shell. Most people who have suffered punishments like Ken Saro Wiwa and others have been as a result of direct government judicial action. More interesting is also the fact that it was Shell that also asked the court to get them out. Was this an attempt at 'guilt-fixing'?

More curious was the fact that the Rossport Five were told that a judge will be on standby day and night to attend to their case if they changed their mind on the contempt. This implies that justice has many faces and can wait for someone. It is important however that governments all over the world have perfected the act of manipulating the judiciary to suit their purpose.

The Rossport Five are lucky to be alive to tell their story. They have even written a book. Willie Corduff won the Goldman Environmental Prize for Europe. In the Niger Delta it is different. Ken Saro Wiwa is dead, Isaac Adaka Boro was executed and what is left behind are tell-tale signs that they have been here. I am still very curious about the fact that in spite of her visibility in the whole Erris saga, Maura Carrington has managed to remain out of trouble. There are so many lessons to be learnt from Maura Carrington.

Chapter Four

Reflections

On the way back from Erris to Dublin to catch a flight, I have casually asked my companion if Shell has done anything good since it came into Erris. My companion paused looked at me and answered, no, nothing good. It struck me that my companion has not seen anything good in Shell, not because as a person and someone from an oil-producing community I have seen much good in oil company activities. It would have been most interesting, if not intriguing to ask a Shell staff if they have done anything good since they came into Erris or to put it in reverse if the community has done anything good since Shell came into the area.

Still on questions, I did ask one of my acquaintances, that since our civilization runs on oil, and that oil and gas pipelines must run through someone's land, farm, compound or community, gas plants must be built on land, what options are there for Shell in Erris. Again he looked at me and said, we do not need it in Erris.

One more final question, I was in Switzerland recently to attend a conference on oil. Almost all the presenters made a show of the presence of China and Malaysia in the international oil market. One human rights activist from Britain was vehement in his presentation about how China does not worry about corporate social responsibility, human rights and good governance.

During my presentation, I pointedly asked the audience if there is anything which Chinese oil companies will do that westerner oil companies have not done. I said that from human rights violations to environmental pollution, from corruption to tax evasion that westerner oil companies and in fact their home governments have all been implicated, so why all the fuss about China?

This is the trend of the discourse in and around extractive industries. And I think that if we keep at it this way, we may never make progress. By the very nature of the oil industry, they operate with a peculiar mindset, whether western or oriental in origin. Shell in Erris is not Chinese, ExxonMobil in Chad is not Malaysian. The issue is that we need oil and need to design better ways of getting oil without incurring too much environmental damage to our ecosystem. It does not matter whether the oil company is Chinese or Dutch, or whether the oil field is at Erris or Umuechem. The point is that there is an increasing threat to the survival of our planet and we need collective and collaborative efforts to save our world.

When I went to the Embassy of Ireland to procure the visa for the trip which culminated in this book, the lady at the counter turned down my request. When I came back her response was that I did not tell her that I have been to Ireland before. I told her that I did not need to tell her that I have been to Ireland that she is supposed to see it from my passport. Second, she would not listen since every Nigerian seems in a hurry to leave the country. And wait for it, the lady at the counter is not Irish (at least by her features), she's Filipino. To her I am like a Chinese oil company that has nothing good to offer. And for once even if I have not been to Ireland, before, there is always a first time. Look at it this way and you will understand her logic. If I have been to Ireland before and did not elope, therefore I cannot elope now. That is exactly how we are responding to the presence of non-western oil companies. If they have not done it before they may not be able to do it now. So if China has been bad in the past, it still has nothing good to offer in the present. The question then is how come China is experiencing this unprecedented growth and every known western nation is in a hurry to invest in China? Can we really say that China has nothing to offer to us in the area of CSR especially as it concerns extractive industries?

When I eventually got to London, my luggages have been sent to Miami. Two things happened. First, I have used shaving powder for about 20 years. Now in Ireland I had the dress on me and my laptop. But I have to spend at least two weeks in Ireland. First, was that I quickly changed to using razor for shaving. I also wore "one and half dress" for almost two weeks. Change came not by design but by default. As I reflect on oil and peacebuilding in our world, the change we seek may come from what John Paul Lederach refers to as seredinpity. Oil company behaviour to my mind may not change by some grand design but by accidental innovations from an innocuous part of the world. Could that be Erris?

When eventually I arrived in Miami, the lady in charge of 'missing' luggage complained that I did not report that my luggage was missing. I said yes, because my luggage was not missing. The Igbos have a saying that something that is where it is kept is not missing. If something is missing, you do not know where it is. But I know that my luggage has been sent to Miami even though I wanted it in Dublin. Since I know where the luggage is, it is therefore not missing. I guess she found my logic weird and did not want to waste her precious time with this queer Nigerian.

Corporate Social Responsibility (CSR) is not missing. It is with us, but we misplaced it. We know where it is and we can go get it and use it. However our fear is that someone may ask us why we did not declare it missing in the first place. It is our lack of candour in explaining that it is not missing that complicates the issue. Every day as I reflect on the concept and theory of CSR, I have always been confronted with the challenge of seeing how CSR was practised in traditional African societies. Though I am aware that the context, time and location may differ, the core ingredients remain essentially the same.

And this reflection has brought me to the inescapable conclusion that for a business to be sustainable, it must be profitable and for a business to be profitable it must address a human problem, because it is in an attempt to solve a human problem that a business generates not only revenue but becomes both profitable and sustainable. But any business that puts profit ahead of customer satisfaction is on its way perdition.

References

Abusharaf, Adila. (1999). The Legal Relationship between Multinational Oil Companies and the Sudan: problems and Prospects. Journal of African Law, Vol. 43, no.1, pp.18-38.

Adewale, Omobolaji (1989). Oil Spill Compensation Claims in Nigeria: Principles, Guidelines and Criteria. Journal of African Law, Vol. 33, No. 1, pp. 91-104.

Atsegbua, L. (1993). Acquisition of Oil Rights Under Contractual Joint Ventures in Nigeria. Journal of African law, Vol. 37, No.1, pp. 10-29.

Azaiki, S. (2003). Inequities in Nigerian Politics. Yenagoa, Nigeria: Treasure Books.

Azam, J.P. (2001). The Redistributive State and Conflicts in Africa. Journal of Peace Research, Vol.38, No.4, pp.429-444.

Azar, Edward (1990). The Management of Protracted Social Conflict. Hampshire, England: Dartmouth.

Bonn International Center for Conversion [Brief 32] (2006). Who's Minding the Store? The Business of Private, Public and Civil Actors in Zones f Conflict. Bonn, Germany: BICC.

Burton, J.W. (ed.) (1990). Conflict: Human Needs Theory. New York: St. Martin's Press

Business Leaders Initiative on Human Rights, Report 2: Work in Progress. London, December 2004.

Catholic Relief Services (June 2003). Bottom of the Barrel: Africa's Oil Boom and the Poor. Baltimore, MD:CRS, p.26

Collier, Paul & Hoeffler, Anke (2001). Greed and Grievance in civil war. Washington, DC: World Bank Publications.

Committee for the Defence of Human Rights [CDHR] (2000). Boiling Point: The Crises in the Oil-Producing Communities in Nigeria. Lagos, Nigeria: CDHR.

Connolly, F., and Lynch, R. (2005). The Great Corrib Gas Controversy. Dublin, Ireland: Centre for Public Inquiry.

Corporate Campaign Inc. (2001) The Arctic Refuge, the Filthy Four and Organized Labor. New York: CC. Inc.

Corrib Gas Update, Issue 2, May 2006

Corrib Natural Gas, A Powerpoint Presentation by Shell (undated).

Crocker, A.C., Hampson, F.O and Aall, P. {eds.}(1996). Managing Global Chaos: Sources of and responses to International Conflict. Washington, DC: USIP.

Denzin, N.K. and Lincoln, S.Y. (eds.) (2003). The Landscape of Qualitative Research: Theories and Issues (2nd edition). Thousand oaks, CA: SAGE Publications.

Druckman, D. (2005). Doing Research: Methods of Inquiry for Conflict Analysis. Thousand Oaks, CA: SAGE Publications.

Falola, Toyin (1999). *The History of Nigeria*. Westport, CT: Greenwood Press.

Frynas, J. G. (2000). Oil in Nigeria: Conflict and Litigation between Oil companies and Village Communities. Hamburg: LIT.

Gary, I. and Reisch, N. (2005). Chad's Oil: Miracle or Mirage? Following the Money in Africa's Newest Petro State. Baltimore, MD: Catholic Relief Services and Bank Information Center.

Gurr, T.R. (1970). Why Men Rebel. Princeton, NJ: Princeton University Press.

Gurr, T.R. (1970). *Why Men Rebel*. Princeton, NJ: Princeton University Press.

Harris, P. & Reilly, B. [eds.] (1998). Democracy and Deep-Rooted Conflict: Option for Negotiators. Stockholm, Sweden: International Institute for Democracy and Electoral Assistance (IDEA).

Imobighe, T.A. et al (2002). Conflict and Instability in the Niger Delta: The Warri case. Ibadan: Spectrum Books.

Interview with a white, adult male, from Erris who prefers anonymity. He is one of the Rossport Five.

Junne, G. & Verkoren, W. {Eds.} (2005). Post Conflict Development, meeting new challenges. Boulder, Colorado: Lynne Rienner Publishers, Inc.

Karl, Terry Lynn. (1997). The Paradox of Plenty: Oil Booms and Petro States (Studies in International Political Economy). Berkeley and Los Angeles, CA: University of California Press.

Kriesberg, L. (2003) Constructive Conflicts: From Escalation to Resolution. Maryland: Rowman and Littlefield Publishers.

Kvale, S. (1996). InterViews: An Introduction to Qualitative Research Interviewing. Thousand, Oaks, CA: SAGE Publications.

Lederach, J.P. (2005). The Moral Imagination: The Art and Soul of Building Peace. New York: Oxford University Press, Inc.

Linstroth, J.P. (2002). History, Tradition, and Memory among the Basques in History and Anthropology, 2002, Vol. 13 (3), pp.159-189.

Manuel, A. (2001). The Dirty Four: The Case Against Letting BP Amoco, ExxonMobil, Chevron and Phillips Petroleum Drill in the Arctic Refuge. Washington, DC: US PIRG.

Marshall, C. & Rossman, G.B. (1999). Designing Qualitative Research {3rd edition}. Thousand Oaks, CA: SAGE Publications.

Murshed, S.M. (2002). Conflict, Civil War and Underdevelopment: An Introduction. Journal of Peace Research, Vol.39, No.4, pp. 387-398.

Okoli, C. (1994). Criminal Liability of Corporations in Nigeria: A Current Perspective. Journal of African Law, Vol.38, no.1, pp.35-45.

Okonmah, P. D. (1997). Right to a Clean Environment: The Case for the People of Oil-producing Communities in the Nigerian Delta. Journal of African Law, Vol. 41, No. 1, pp. 43-67.

Okonta, Ike and Oronto Douglas (2001). *Where Vultures Feast*. San Francisco, CA: Sierra Club Books.

Omeje, Kenneth. (2006). Petrobusiness and Security Threats in the Niger Delta, Nigeria. Current Sociology, Vol. 54 (3): 477-499, May 2006. Sage (London, Thousand Oaks, CA and New Delhi).

Onuoha, A. (2005). From Conflict to Collaboration; Building Peace in Nigeria's Oil-Producing Communities. London: Adonis & Abbey Publishers, Ltd.

Onuoha, A. (2007). The Dilemma of Restorative Justice when "All are Guilty:" A Case Study of the Conflicts in the Niger Delta Region of Nigeria. African Journal on Conflict Resolution, Volume 7, Number 1, 2007.

Osaghae, E. E (1995). The Ogoni Uprising: Oil Politics, Minority Agitation and the Future of the Nigerian State. *African Affairs*, Vol. 94, pp. 325-344.

Osaghae, E. E. (1991). Ethnic Minorities and Federalism in Nigeria. *African Affairs*, Vol. 90, No. 359, pp.237-258.

Osaghae, E. E. (1998). Managing Multiple Minority Problems in a Divided Society: Nigerian Experience. Journal of Modern African Studies, 36 (1), pp. 1-24.

Ramsbotham, O., Woodhouse, T., and Miall, H. (2005). Contemporary Conflict Resolution (second edition). Cambridge, UK: Polity Press.

Ritzer, G. and Douglas J. Goodman (2004). *Sociological Theory* (6th Edition). New York: McGraw Hill.

Ross, M.H. (1993). The Management of Conflict: Interpretations and Interests in Comparative Perspective. New Haven: Yale University Press.

Ross, Michael (2001). How does natural resource wealth influence civil war? Los Angeles, CA: University of California Press.

Sharp, G. (2005). Waging Nonviolent Struggle; 20th century Practice and 21st century potential. Boston, MA: Porter Sargent Publishers, Inc.

Shell Nigeria 2004 Annual Report

Stakeholder Research Associates (2005). From Words to Action: The Stakeholder Engagement Manual. Volume 1: The Guide to Practitioner's Perspectives on Stakeholder Engagement. Canada: Stakeholder Research Associates.

Staub, E. (2003). The Psychology of Good and Evil: Why Children, Adults, and Groups Help and Harm Others. Cambridge, U.K. and New York: Cambridge University Press.

Traub-Merz, R. & Yates, D. [eds.] (2004). Oil policy in the Gulf of Guinea: Security & Conflict, Economic Growth, Social Development. Bonn, Germany: Friedrich-Ebert-Stiftung.

USAID Office of Conflict Management and Mitigation. (2004). Land & Conflict; A   Toolkit for Intervention. Washington, DC: USAID.

Wokoma, I.N. (undated). Assessing Accomplishments of Women's Nonviolent Direct Action in the Niger Delta. (no publisher).

Woolfson, Charles and Beck, Mathias (eds.). (2005) Corporate Social   Responsibility Failures in the Oil Industry (Work, Health and Environment). Amityville, New York: Baywood Publishing Company, Inc.

www.shelltosea.org

Index

100

*9 7 8 1 9 0 6 7 0 4 0 4 9 *